SOLAR SYSTEM
OBSERVER'S GUIDE

Peter Grego

FIREFLY BOOKS

Published by Firefly Books Ltd. 2006

First printing

Publisher Cataloging-in-Publication Data (U.S.)

Grego, Peter.
Solar system observer's guide / Peter Grego.
[256] p. : col. ill., photos., maps ; cm.
Includes bibliographical references and index.
Summary: An introduction to observing the solar system with binoculars or small telescopes. Includes information on the moon, sun, meteors, comets, asteroids and nine planets.
ISBN 1-55407-132-1 (pbk.)
1. Astronomy — Observers' manuals. 2. Solar system — Popular works. I. Title.
523.2 dc22 QB501.G74 2006

Library and Archives Canada Cataloguing in Publication

Grego, Peter
Solar system observers guide / Peter Grego.
Includes bibliographical references and index.
ISBN 1-55407-132-1
1. Astronomy—Observers' manuals. I. Title.
QB501.2.G74 2006 523 C2006-904369-5

Published in the United States by
Firefly Books (U.S.) Inc.
P.O. Box 1338, Ellicott Station
Buffalo, New York 14205

Published in Canada by
Firefly Books Ltd.
66 Leek Crescent
Richmond Hill, Ontario L4B 1H1

ISBN 13: 978-1-55407-132-6

Printed in China

To my mother, Margaret Joan Grego, and my late father, Terence John Grego, who encouraged me to take my first tentative steps out into the Solar System.

Front Cover: *(above left) Neptune (NASA/STScI). (below right) Triton (NASA/JPL/Univ. of Arizona). In the foreground is an 8-inch Newtonian reflector on an equatorial mount (Jamie Cooper).*

Back Cover: *(above) Martian volcano Olympus Mons (NASA/JPL). (below) Orbits of Mercury and Earth, to scale, in early 2007 (Peter Grego/Philip's).*

Title Page: *Jupiter from the Cassini spacecraft (NASA/JPL/Univ. of Arizona).*

CONTENTS

1 · OBSERVING THE SOLAR SYSTEM

The Solar System, our cosmic backyard, comprises the Sun and everything within its gravitational domain. This includes all the major planets and their satellites, the asteroids and comets. Far and away the brightest and most easily accessible objects to the observer, the Sun and the Moon are spectacular when viewed through a small telescope (the Sun requires special precautions to view safely – see Chapter 15 for details on safe solar observing). Though they're much smaller and fainter, the brighter planets can also appear magnificent when viewed through the telescope eyepiece. Each planet displays its own unique set of phenomena. Spectacular phenomena visible in the Earth's atmosphere, such as aurorae and meteors, are best viewed with the unaided eye. Binoculars show bright comets well, often revealing considerable detail within their wispy tails.

Wherever you live, each clear night there's no shortage of opportunities to observe a variety of objects and phenomena within the Solar System, whether you're using the unaided eye, binoculars, or a telescope of any sort, small or large. As a whole, Solar System observation is far less hampered than deep-sky observation by the detrimental effects of light pollution, be it the type inflicted by thoughtless neighbors or the miasma blanketing whole urban regions.

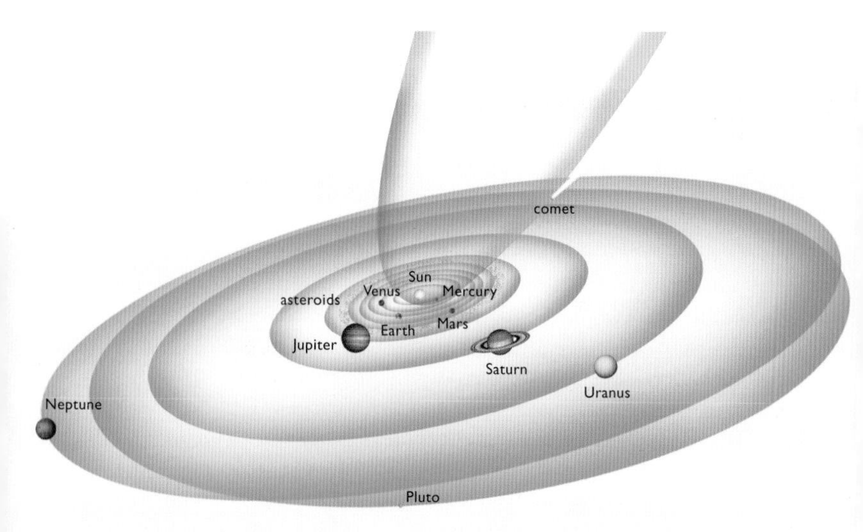

▲ *A diagram showing the Solar System from around 30° above the general plane of its major planets. It is not to scale.*

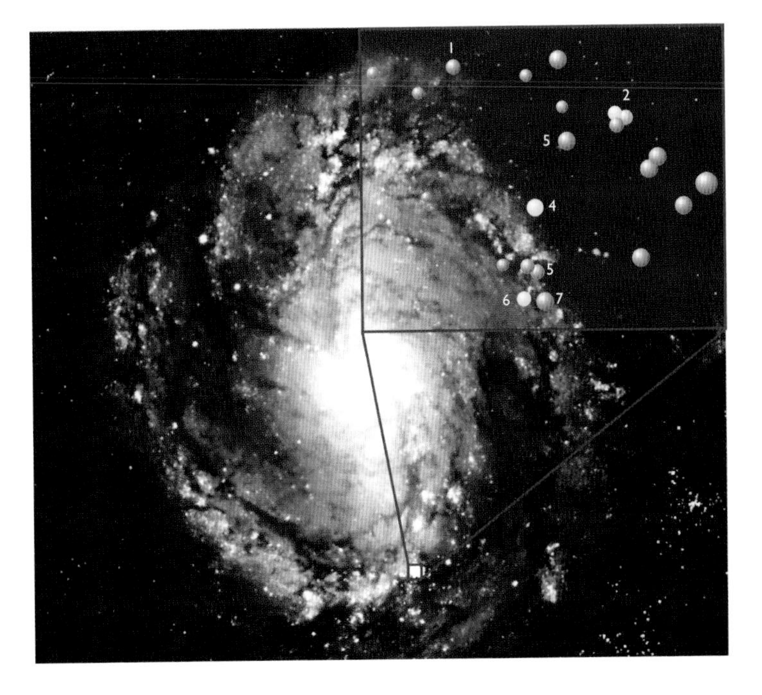

▲ *The Solar System's place within the Milky Way, and our local stellar neighborhood (not to scale). 1 Barnard's star; 2 Alpha Centauri; 3 LP387-541; 4 Sun; 5 UV Ceti; 6 Tau Ceti; 7 Lal 21185LP387-541; 4 Sun; 5 UV Ceti; 6 Tau Ceti; 7 Lal 21185.*

Our cosmic backyard

In an outlying part of the Milky Way, around 25,000 light years from the Galactic center, lies a star called the Sun. It's a middling star, of average mass and brightness. There are hundreds of millions of stars just like it in the Milky Way. It takes the Sun around 226 million years to make one orbit around the Galactic center. Within a single Galactic orbit since the Sun's birth, the dusty, gassy remnants of the material surrounding the Sun flattened into a disk, and planets formed within it. Such a process is not unique, or even rare – disks of planet-forming material are thought to develop around most newborn stars. The biggest of these newly formed planets, Jupiter, failed to sweep up enough gas and dust for its core to become dense enough and hot enough to shine as a star in its own right. It did, however, accumulate enough material to become an imposing ball of gas one-tenth the diameter of the Sun. Three other large gas giants – Saturn, Uranus and Neptune – formed further away from the Sun. Nearer the Sun, four much smaller planets formed – Mercury, Venus, Earth and Mars – each having a solid, rocky surface. A curious thing happened on the third rock during the Sun's second orbit around the Galactic center – life appeared. Within the space of another dozen orbits of the Galactic center, an evolutionary branch of terrestrial life had developed into a race of sentient beings capable of viewing the Universe around them – from the Moon to the furthest reaches of the cosmos – with wonder and curiosity.

Although the Solar System, our cosmic backyard, is but a tiny place in the cosmic scheme of things, it is of the utmost significance to its inhabitants. The Earth is the only place we know where life exists, and the Sun is the only star that we can study in detail. Each major planet in the Solar System is unique. Each has its own individual surface conditions and its own peculiar type of atmosphere. While all of the solid worlds share a wide variety of geological processes, each possesses its own unique rocky make-up and its own set of unique surface features. The gas giants are all remarkably individual, too, each having a different atmospheric cocktail and each displaying a wide variety of colors and atmospheric activity.

Many moons

With the exception of Mercury and Venus, all the planets have satellites orbiting them. Most of them are pretty small in comparison with the planets they revolve around. Among Jupiter's more than 60 known satellites, four of them – Io, Europa, Ganymede and Callisto, known as the "Galilean moons" after their discoverer, Galileo – are exceptionally large, planet-sized objects. They are, in fact, big enough to be easily visible through a pair of binoculars, shining as bright star-like points of light near the dazzling Jovian disk. Yet these big satellites of Jupiter are tiny in comparison with Jupiter itself, and so are most other planetary satellites.

Earth's only natural satellite, the Moon, measures one-quarter the Earth's diameter, and is particularly large in comparison. Charon, Pluto's only known satellite, is comparatively even larger, and astronomers often refer to Pluto and Charon as a double planet. The Moon and Charon may share the same mode of origin, both having possibly been sliced off their respective parent planets in the remote past by the impact of another sizable planetary body. The Galilean moons, and the large satellites of Saturn, Uranus and Neptune, are likely to have been formed by gravitational accretion in orbit around their parent planets. Many of the smaller satellites in the Solar System, like diminutive Phobos and Deimos orbiting Mars, and the hundreds of irregular fragments orbiting the gas giants, are probably captured asteroids and cometary nuclei. Each of the four gas giants has a ring system made up of countless trillions of small orbiting moonlets ranging from grains of dust to house-sized chunks of rock and ice. Saturn's rings are the only ones visible through the eyepiece of a backyard telescope.

Interplanetary debris

There's much more to the Solar System than the major planets and their satellites. Judging by the rate at which new asteroids and comets are being discovered, there are likely to be millions of them larger than a kilometer in diameter in orbit around the Sun. Hundreds of thousands of asteroids (or "minor planets") orbit the Sun within the main asteroid belt, between the orbits of Mars and Jupiter. The asteroids are spread throughout such a large volume of space that if you were standing on one, the nearest one would likely appear as a distant, dim star-like point. In addition to the minor planets within the main asteroid belt, there are other distinct groups of asteroids elsewhere in the Solar System. Near-Earth asteroids are those whose orbits take them close to that of the Earth, including potentially hazardous asteroids – objects that could possibly collide with us at some point in the distant future. An interesting group called the Trojan asteroids are clustered at points far preceding and following Jupiter in its orbit around the Sun, locked there in a gravitational resonance.

It is thought that many of the asteroids in the main belt were part of larger bodies that broke up. These parent bodies, formed early in the history of the Solar System, grew hot inside and developed a core, mantle and crust. Most of the meteorites found on the Earth are fragments of asteroids, and by studying their composition we can tell what the original parent bodies were made of.

Comets frequently glide into the inner Solar System and brighten enough to be visible through binoculars or with the unaided eye. When far away from the Sun,

▶ *Easily visible with the unaided eye, Comet Hale–Bopp made a splendid appearance in 1997. It is seen here in an image taken by Jamie Cooper.*

chilling out in the interplanetary deep freeze, a comet is a pretty unimpressive sight – a solid ball of ices mixed with dirt and rock measuring just a few kilometers across. Known as the nucleus, this "dirty snowball" is made up of material left over from the formation of the Solar System. As it approaches the Sun, the comet's nucleus heats up and its ices sublimate, turning directly from a solid state into gas. Streaming off the nucleus, this gas carries with it grains of dust, forming a large coma shining by reflected and scattered sunlight, and it may go on to develop a prominent tail, perhaps spanning interplanetary scales.

Halley's Comet, the most famous cometary visitor, is just one of more than 160 periodic comets known to make regular visits to the inner Solar System. Halley's orbit takes it from the frigid realms beyond Neptune to the inner Solar System every 76 years. Some comets, such as the spectacular Hale–Bopp of 1997, have orbital periods of many thousands of years. Another type of comet has never been warmed by the Sun before, having spent all its life at the very limits of the Sun's influence, in a region known as the Oort Cloud. Thought to be a vast shell made up of billions of cometary nuclei, the Oort Cloud is so vast that it extends half-way to the nearest stars.

Comets leave behind trails of dusty debris in space. Sometimes the Earth plows through this material, and each particle entering the Earth's atmosphere quickly burns up and appears as a meteor. There are several bright meteor showers visible on regular, predictable dates each year, and some of these have had their parent comets identified. Each annual shower appears to emanate from a specific point in the sky known as a radiant, and the shower is named for the position of the radiant – for example, October's Orionid meteors have their radiant in the constellation of Orion, and their parent is Halley's Comet.

Planetary phenomena

As the Earth orbits the Sun, the Sun appears to trace a path against the background constellations throughout the year. This path – the Earth's orbital plane – is called the ecliptic. All the planets, with the exception of Pluto, have orbital planes roughly coinciding with that of the Earth, and they therefore all follow paths within a few degrees of the ecliptic plane. The Moon's orbital plane around the Earth lies close to the ecliptic, too. Being a path common to the Sun, Moon and planets, the ecliptic is a busy thoroughfare. The Moon occasionally passes in front of the Sun, producing solar eclipses. The Moon and planets often appear to approach each other very closely. Sometimes the Moon moves directly in front of a planet, producing an occultation. Close approaches between planets are called appulses, and conjunctions occur when two planets share the same Right Ascension. On

▶ *Mars (lower left) and Saturn (lower middle) lie a few degrees south of the crescent Moon, which is adorned with earthshine in this image by Jamie Cooper.*

very rare occasions one planet will appear to move in front of another, producing a mutual occultation; the next one will take place in 2123, when Venus passes in front of Jupiter. The Sun also regularly appears to make close approaches to the planets, but the solar glare renders such events unobservable, apart from those occasions when Mercury and Venus transit the Sun (as explained below).

Phenomena of the inferior planets

Since the orbits of Mercury and Venus both lie within the Earth's orbit, they are referred to as the inferior planets. From our vantage point, neither planet appears to stray very far from the Sun. Both Mercury and Venus display a sequence of phases during each elongation from the Sun, in addition to changes in their apparent diameter, which can be followed through a small telescope. Following an inferior planet from its superior conjunction, when it lies directly on the far side of its orbit, hidden from view by the Sun's glare, the planet begins to edge east of the Sun. Once it has moved far enough away from the Sun to be seen against a reasonably dark evening sky, a telescope will reveal it as a small gibbous disk. The planet's phase lessens as it moves further away from the Sun, until it reaches half phase, or dichotomy, at its greatest elongation east of the Sun. The planet then begins to move toward the Sun, its phase becoming a large crescent, until it is lost in the Sun's glare once more. Inferior conjunction is reached when Mercury or Venus passes between the Earth and the Sun. On most occasions the planet will pass some distance to the north or south of the Sun at inferior conjunction, but on those rare occasions when an inferior planet happens to pass exactly between the Earth and Sun, it appears to transit the Sun's disk and can be seen as a small circular silhouette. As the planet draws to the west of the Sun, it eventually becomes visible in the predawn skies, when telescopes will reveal it to be a large crescent phase. The planet's phase gradually fills

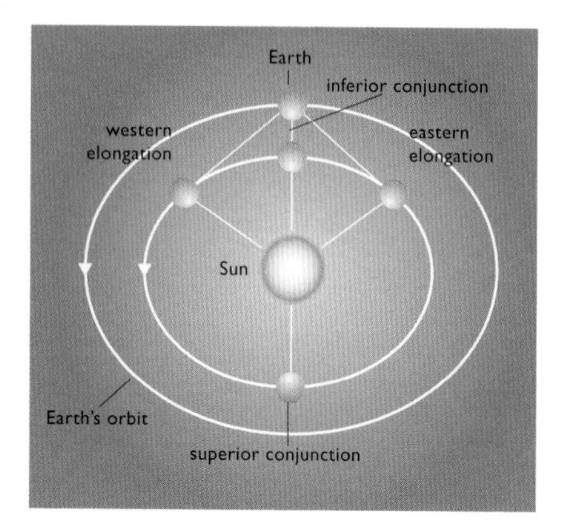

◀ *Orbital phenomena of the inferior planets.*

out, and its apparent diameter slowly decreases. Dichotomy is reached at its greatest elongation west of the Sun. As the planet draws closer to the Sun once more, it becomes a more gibbous phase, growing ever smaller, until it is lost in the Sun's glare. The entire sequence begins again at superior conjunction.

Not all morning and evening apparitions of Mercury and Venus are favorable. From temperate latitudes, the angle made by the ecliptic at the sunset or sunrise horizon varies considerably from season to season. In the northern hemisphere, eastern elongations of Mercury and Venus are most favorable during the spring months, when the planets are highest above the western horizon as the Sun sets. Western elongations of the inferior planets are most favorably observed during the autumn, when they are at their highest above the eastern horizon before sunrise.

Phenomena of the superior planets

Mars, Jupiter, Saturn, Uranus, Neptune and Pluto are collectively known as the superior planets because they all orbit the Sun (in the order given) outside the orbit of the Earth. Conjunction with the Sun takes place when a superior planet is on the far side of the Sun from the Earth. Since the Earth is on a faster orbital circuit than the superior planets, they appear to edge west of the Sun after conjunction (the opposite direction to the inferior planets, which both move east of the Sun after superior conjunction). As they clear the glare of the Sun, the superior planets begin to peek out of the predawn skies. At this stage the superior planets have their smallest apparent diameter.

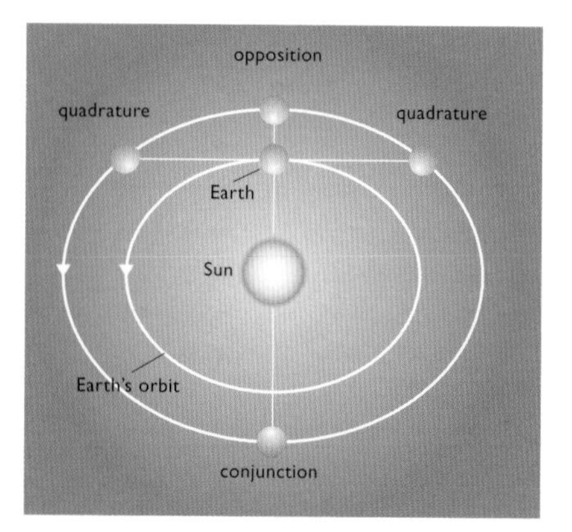

◀ *Orbital phenomena of the superior planets.*

Jupiter and Saturn both have apparent diameters that are easily large enough for serious telescopic observation to commence as soon as they become visible against a reasonably dark sky at a sensible altitude above the predawn horizon. Jupiter always appears larger than 30 arcseconds, and the apparent diameter of Saturn's disk always exceeds 18 arcseconds. Mars, however, only becomes a visually interesting object when it grows larger than 5 arcseconds in apparent diameter, when its ice caps and broader desert markings can just about be discerned through a 100 mm telescope. The magical 5 arcseconds apparent diameter is reached many months after Mars first becomes visible with the unaided eye in the morning skies. For example, Mars was at conjunction on September 15, 2004, yet its apparent diameter had grown to 5 arcseconds only by mid-February 2005, some five months later. Mars had grown to 10 arcseconds by mid-July 2005, and reached opposition on November 7, 2005, when it was 20 arcseconds across.

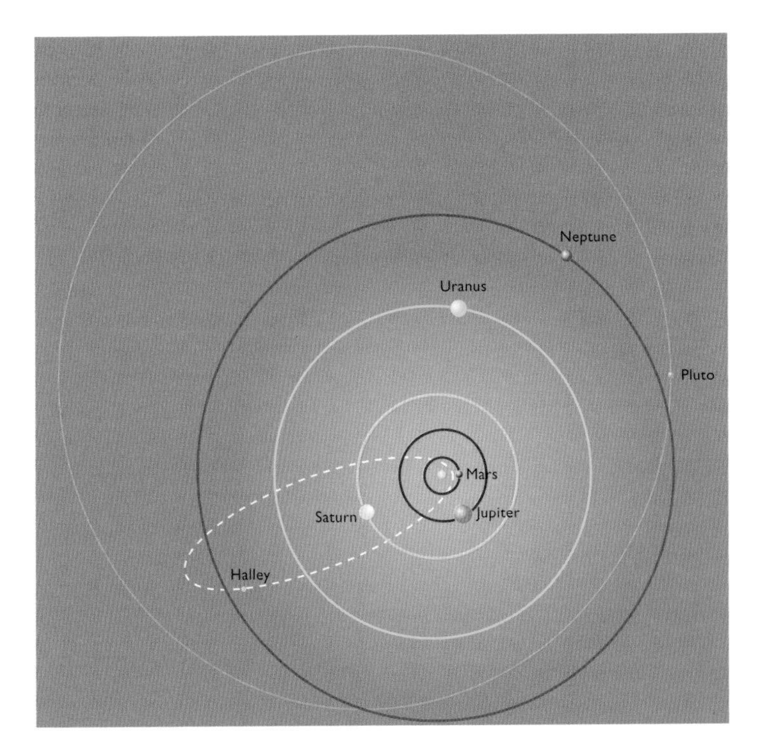

▲ *Orbits of the outer planets, shown to scale. The orbits of Halley's comet (dashed line) and Pluto are strongly inclined to the ecliptic, so there is no danger of collision where their orbits appear to intersect those of the other planets.*

At opposition, each superior planet is located opposite the Sun in the sky, and it is due south at midnight. The superior planets are virtually 100% illuminated at opposition and are at their largest apparent diameter for that particular apparition. Since each planet (including the Earth) has a slightly elliptical orbit, the distance between the Earth and the superior planet varies at opposition. Mars displays by far the greatest variation in opposition diameter, ranging from a minimum of around 15 arcseconds at aphelic oppositions (when it is furthest from the Sun) to 25 arcseconds at perihelic oppositions (when it is nearest the Sun).

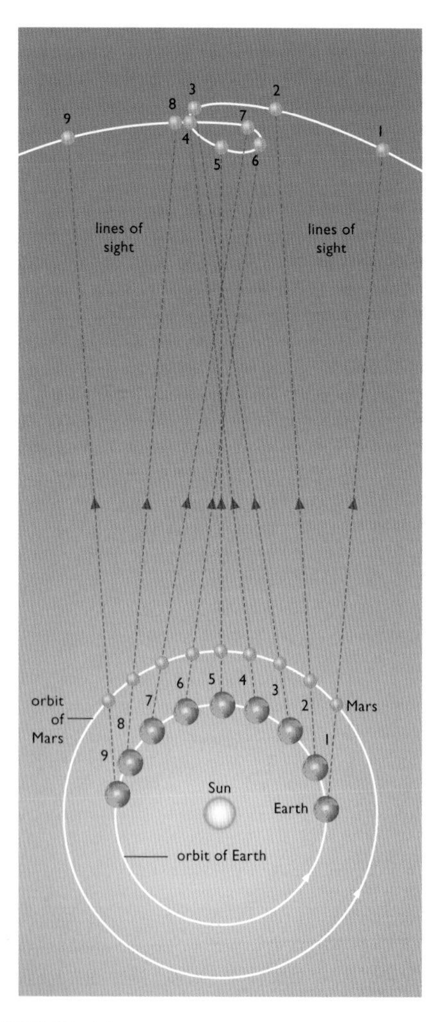

Of course, the slow drift of the superior planets to the west of the Sun during their apparitions is caused by the movement of the Earth around the Sun. Although the aggregate movement of the superior planets against the celestial background is slowly eastward, a phenomenon called retrograde motion causes a superior planet to reverse direction against the background stars for a while, the planet performing a small loop or a zigzag in the sky before proceeding eastward once more. Beginning some months prior to a planet's opposition and ending some months afterward, retrograde motion is caused by our view of the planet from the Earth and our shifting line of sight. Following conjunction with the Sun, a superior planet appears to move sedately to the east. As the Earth, moving along its faster inner circuit, catches up with the

◀ *Superior planets display retrograde motion with respect to the stars during each apparition. It is caused by our changing line of sight as the Earth "overtakes" the planet on its inside track around the Sun.*

planet, prior to its opposition, our moving line of sight causes the planet's apparent motion to slow down and then move westward for a while. As we draw further away from the planet after opposition, our line of sight begins to alter the apparent course of the planet; it appears to slow down and then finally recommences its slow eastward path against the stars. The size of the retrograde paths made against the background stars lessens with the distance of the planet. Minor planets, orbiting in the main asteroid belt between Mars and Jupiter, also display retrograde motion during the course of an apparition.

Observing the planets

All the major planets are capable of being viewed by the well-equipped amateur astronomer. Five planets are bright enough to be seen with the unaided eye in a reasonably dark sky. Mercury, Venus, Mars, Jupiter and Saturn all have such large apparent diameters that their planetary nature becomes evident when they are viewed at a medium magnification through a small telescope. Uranus and Neptune can easily be picked up through binoculars, but their apparent diameters are so small that a fairly high magnification is required to see them as distinct disks. Pluto (discovered as recently as 1930) can be located through a medium-sized amateur telescope, but its disk cannot be resolved visually.

Elusive Mercury

Mercury, the planet nearest the Sun, is the most elusive of the five naked-eye planets. It completes each orbit around the Sun in a little under 88 days, and since it never strays more than 30° from the Sun, naked-eye observers have only a few days to spot Mercury each apparition. Observers need as clear and unpolluted a horizon as possible in order to identify Mercury. As the sky darkens, Mercury appears to brighten. During each apparition, Mercury is at its brightest when it is about 80% illuminated, sometimes exceeding the apparent brightness of the star Arcturus. Some observers have noted a distinct pinkish hue, but this is probably an artifact caused by the Earth's atmosphere. When using binoculars to spot Mercury in the evening, the observer should wait until the Sun has safely set before beginning to sweep the horizon. Even a brief flash of direct sunlight transmitted through any optical equipment to an unprotected eye can cause discomfort, eye damage and possibly permanent blindness.

Through a telescope Mercury presents a rather small disk. Its phase is noticeable at higher magnifications, but surface markings are elusive. Many observers choose to observe Mercury when it is high in the sky during full daylight, either by continuing to follow the planet after sunrise, or by using setting circles or an electronic go-to system to locate

▲ *The crescent Moon and Venus, captured by Jamie Cooper.*

it. It is futile, and potentially dangerous to one's eyesight, to attempt to locate Mercury (or for that matter any other object near to the Sun) with a telescope during the daytime by sweeping the skies in a haphazard fashion.

Dazzling Venus

Venus appears as a brilliant white star during its morning and evening apparitions, ranging between magnitude −3.5 and −4.3. At greatest elongations Venus can stray up to 47° away from the Sun – far enough, during favorable elongations, to place it high in a dark sky for several hours of observation. It is possible to view Venus in full daylight with the unaided eye, provided it's far enough from the Sun and you know exactly where to look. Binoculars are sufficient to reveal the phase of Venus when it is a sizable crescent, and a telescope will show a sequence of phases during an apparition. Venus is smothered in thick clouds and its surface can never be seen visually. At first sight through the telescope eyepiece, Venus presents a blank, brilliant disk. Optical effects produce all manner of spurious artifacts, such as rays, spikes or ghosting – every telescope is prone to them, to some degree, so don't be discouraged if the image doesn't look picture-perfect. Once the observer gets used to the planet's glare, careful observation is likely to reveal traces of atmospheric detail, often taking the form of broad, curved dusky swathes.

Mars – worth waiting for

Mars reaches opposition every 780 days or so, during which time it appears as a brilliant star with a distinctly orange hue, but it can be easily followed with the naked eye for 18 months each apparition. Mars has a more eccentric orbit than the Earth, so not all oppositions are equally as good. Telescopes show an orange disk with dusky surface markings and bright polar ice caps. Occasional bright cloud features and yellow dust storms are visible.

Minor planets

Binoculars will reveal several of the brighter minor planets which orbit in the main asteroid belt between Mars and Jupiter. Dozens of minor planets can be viewed through a 100 mm telescope – indeed, every clear night presents an opportunity to locate minor planets telescopically. Minor planets are so small and distant that none of them can be resolved visually as disks – they simply appear as points of light, hence the name "asteroid" (Greek: star-like). It is an enjoyable exercise to locate the brighter minor planets and to plot their course over a period of days, weeks or months on a star chart.

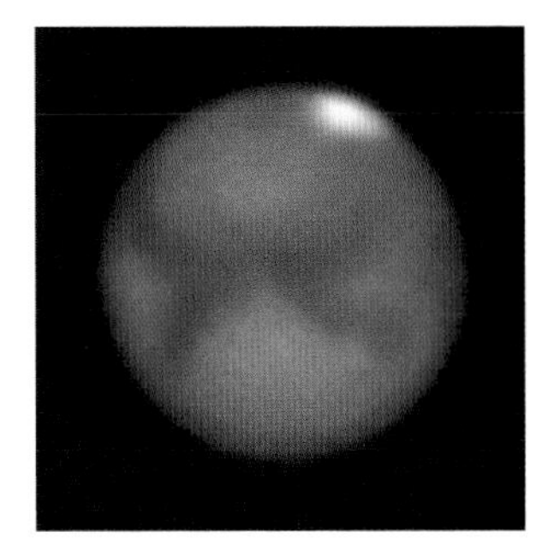

▲ *Detail on Mars – its dusky deserts and bright polar cap – can easily be viewed when it is near opposition. This image, taken by Jamie Cooper, shows Mars near its close perihelic opposition in August 2003.*

Tumultuous Jupiter

To the unaided eye, Jupiter appears as a brilliant white star, sometimes shining as bright as magnitude −2.9. Of all the planets, Jupiter offers the most dynamic spectacle – a changing scene that can be followed and enjoyed through modest instruments and studied in great detail by careful, methodical observation.

A small, steadily held pair of binoculars will show Jupiter's disk and its four bright Galilean moons, Io, Europa, Ganymede and Callisto. Jupiter, a rapidly spinning gas giant, has a noticeably flattened disk and is crossed by a number of bright and dark cloud belts. The intensity of these belts varies over the years, but the widest and most prominent dark ones are usually the two equatorial belts. Closer telescopic scrutiny reveals that Jupiter's atmosphere contains numerous spots, ovals and festoons. Jupiter's most famous feature, the Great Red Spot, is a giant anticyclone that has been active for centuries. Variations in the color and intensity of the Great Red Spot occur over time – sometimes it fades into obscurity, leaving only a hollow in the South Equatorial Belt to indicate its presence. At

▼ *Jupiter is the largest planet in the Solar System. Features in its turbulent atmosphere can be viewed through small telescopes, including the Great Red Spot, which is visible in this image taken by Jamie Cooper.*

other times, it has been so dark and intense as to have been described as a deep "brick red" color.

Phenomena of the Galilean moons are fascinating to follow through a telescope. These planet-sized satellites can be discerned as tiny disks through a 100 mm telescope. As they orbit Jupiter in the plane of its equator, the Galileans occasionally appear to move directly across the face of Jupiter, preceded by, or trailing their shadows. Frequently eclipsed by the shadow of Jupiter, the satellites play a continual game of hide and seek as they peek out from, or slip behind Jupiter's limb. Sometimes, mutual occultations and eclipses between Jupiter's moons take place.

Stunning Saturn

Saturn appears as a yellowish star, always shining brighter than magnitude 0.9. Apart from those occasions when Saturn's rings are presented edge-on to the Earth, it is obvious even through binoculars that Saturn doesn't have a simple circular shape. A small telescope will reveal a flattened cream-colored disk surrounded by a glorious ring system, and higher magnifications show the Cassini Division in the rings, the shadow of Saturn's globe on the rings and the shadow of the rings on the globe. Titan, Saturn's biggest satellite, is visible in binoculars, and telescopes will show several additional satellites.

Saturn is crossed by dusky atmospheric belts lying parallel to its equator, but they are far less obvious than those of Jupiter. Activity within Saturn's atmosphere is not often very obvious, and features such as prominent dark or bright spots are rare. Large bright spots suddenly appeared on Saturn in 1933, 1960 and 1990 – each was fascinating to follow, but all faded from prominence after a few weeks.

◀ *Saturn never fails to impress first-time and veteran viewers alike. The image was taken by Jamie Cooper.*

The outer limits

At its brightest, Uranus is just about visible to a keen-sighted observer when it is reasonably high in dark, transparent skies. Neptune, the most distant gas giant, never shines brighter than magnitude 7.6 and it requires optical aid to locate. Viewed through the telescope eyepiece, Uranus and Neptune present rather small disks. At its best, Uranus almost reaches 4 arcseconds in apparent diameter, and its pale green disk is quite easy to perceive through a small telescope using a high magnification. Neptune only just exceeds 2.5 arcseconds in apparent diameter at its best, and its blue-tinted disk presents a challenge to discern through a small telescope. Binoculars will best show the colors of both planets.

Icy Pluto – the smallest planet in the Solar System – has the most eccentric orbit of all the planets, and the one most steeply inclined to the plane of the ecliptic. Pluto appears as a faint star-like point, shining feebly at a maximum magnitude of 13.6. Viewed through the eyepiece of the largest amateur telescope, Pluto will not reveal its disk.

Deep frozen worlds

Around 1000 sizable icy worlds have been discovered orbiting the Sun beyond Neptune. In July 2005 the discovery of an object substantially larger than Pluto, designated 2003 UB313, was announced. Unofficially known as "Xena," it orbits the Sun every 560 years at a distance of between 38 and 97 AU (one AU, an Astronomical Unit, is the distance between the Earth and Sun, some 150 million km). Many more sizable distant worlds are thought to lurk in the far reaches of the Solar System.

Lunar landscapes

Lunar observation is one of the most visually rewarding branches of astronomy. Viewed without optical aid the illuminated face of the Moon appears as a patchwork of dark and light areas. One-third of the Moon's near-side is covered with maria (Latin: "seas"), which appear as dark patches. Although the earliest lunar observers may once have imagined the maria to be tracts of water, astronomers have known for centuries that they are solid plains. We now know that they are made of lava that flowed across the floors of large asteroid impact basins several billion years ago.

▲ *The young crescent Moon, captured by Jamie Cooper.*

Observed through a modest pair of binoculars, the Moon's surface resolves into a remarkable collection of seas, mountains and craters. The brighter areas surrounding the maria are shown to be mountainous, highly cratered regions. Here and there, radiating from a number of bright craters, light-colored rays streak the Moon's surface.

A wealth of lunar detail is visible in the eyepiece of a small telescope. Thousands of craters can be seen, ranging from big lava-flooded craters, sometimes called "walled plains," to prominent asteroid impact craters with their bright ray systems. Just about every other class of lunar feature is discernable – mountains, hills, domes, rilles, clefts, faults and valleys.

Solar splendors

The surface of our nearest star is in constant turmoil. A safely used small telescope reveals great detail. Dark spots develop on the Sun's glowing surface, and their movement across the Sun's face can be followed on a day-to-day basis, along with changes in their size and complexity. Most sunspots have a lifetime of less than one rotation of the Sun (around 25 days at the Sun's equator), but large ones can sometimes be followed for several months. Under good conditions, large telescopes reveal a fine granulation of the Sun's surface, caused by bubbling convection cells. Brighter areas called faculae can sometimes be seen, usually appearing more prominent when near the Sun's edge.

It is dangerous to view the Sun directly through binoculars or a telescope – a fraction of a second's glimpse of magnified sunlight is sufficient to cause permanent eye damage, even blindness. The safest way to observe the Sun is to project its image onto a shielded smooth white card. If your telescope has a finder, keep its lenses covered and never attempt to locate the Sun with it. Never use small dark filters that fit on eyepieces, and never use any other household materials as a solar filter – such as a CD-ROM – as these are utterly inadequate to protect your eyes from the potentially blinding solar radiation. More experienced observers use special solar filters that fit over the telescope's entire aperture.

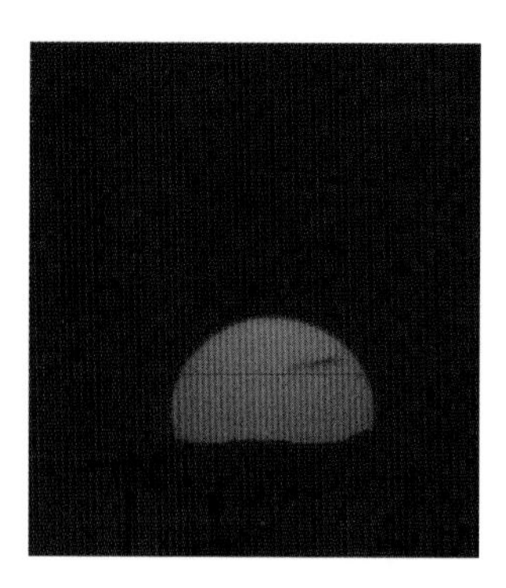

▲ *Sunset captured by Jamie Cooper.*

A whole Solar System to explore

This has been just a brief introduction to the joys of observing the Solar System. Within the keen observer's grasp is a host of amazing worlds, a never-ending parade of remarkable visual phenomena and enough variety and surprise to keep anyone enthralled for a lifetime. From the blazing Sun to deep-frozen Pluto, within a volume of space ten light hours across, all these worlds are yours to explore.

2 · THE SOLAR SYSTEM EXPLORER'S TOOLS

Vision

Descriptions of the optical configuration of various types of binoculars, telescopes, eyepieces and accessories appear in most general astronomy books, so it's not surprising that most amateur observers have a pretty sound knowledge of the optical equipment that they use to view the skies. Every amateur astronomer can tell the difference between a Galilean refractor and a Newtonian reflector; most can explain why a Nagler eyepiece is optically superior to a Huygenian eyepiece; and some optical aficionados can confidently describe the merits of a Shiefspiegler for advanced planetary observation. The advantages of a number of commonly used telescopes, eyepieces and accessories for Solar System observing are described later in this chapter.

Surprisingly few amateur astronomers blessed with good vision pay close attention to the most important optical equipment of all – their own eyes. By knowing about the structure of the eye, how the eye works, and how our brains perceive visual information, one's enjoyment of visual observation can be tremendously enhanced.

How the eye works

Often compared to the workings of a camera, the human eye is in fact a far more complex and flexible piece of biological "engineering" than the most sophisticated digital camera on the market today. The eye consists of two fluid-filled compartments – the aqueous humor at the front of the eye and the larger vitreous humor, making up the bulk of the eyeball. Incoming light passes through the cornea, a transparent

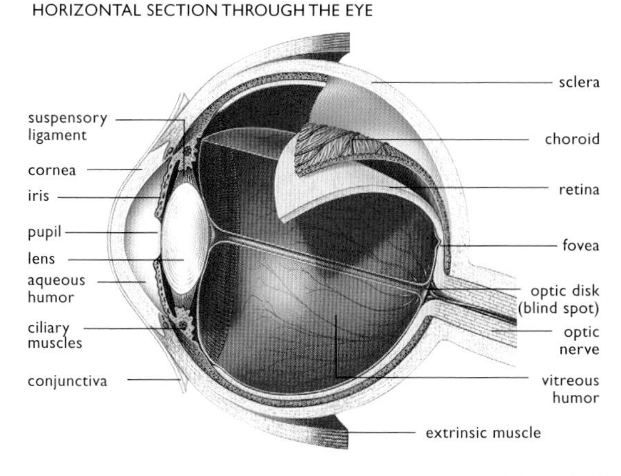

► *Highly sophisticated and eminently adaptable, the human eye has a dynamic range far greater than any CCD camera.*

membrane covering the front of the eye, through the aqueous humor, then through a crystalline lens behind the pupil, across the vitreous humor, to finally impact onto the retina on the inside surface of the back of the eyeball.

Both the cornea and the lens serve to focus the light, with the moment-by-moment fine-tuning being performed solely by the lens. The amount of light passing through the lens is regulated by the iris, whose fine muscular fibers alter the diameter of the pupil, the aperture at its center. Light-sensitive cells in the retina convert the image – which, incidentally, has been turned upside-down by the lens – into electrical pulses which are sent to the brain and processed into an image which we perceive as being the right way up.

Vision in focus

Our entire field of view spans almost 180° horizontally and around half this vertically. Out of this wide vista, the extent of our field of distinct vision, where everything appears in sharp focus, is only around two degrees. When we look up at the night sky, we can only see a few stars at a time in sharp focus. Yet our perception is very different. Unable to rest upon a single static object and hold it at the center of our vision for any length of time at leisure, our eyes are continually scanning a scene to accumulate the maximum amount of information from it.

This physiological phenomenon has direct consequences on our attempts to observe objects with a small apparent diameter. For example, the Moon has an apparent diameter of around half a degree when viewed with the unaided eye – easily small enough to be hidden by the tip of the little finger. On cursory examination, the Moon's patchwork of light and dark areas can be seen without any problem by anyone with reasonably good vision. It can be challenging, however, to make an accurate observational drawing of the detail visible on the lunar disk with the unaided eye. While the broad lunar features are easy enough to place, when it comes to satisfactorily recording fainter, smaller features, it can be frustrating. The impression most people get is that there is plenty of fine detail, but much of it cannot be held steadily in view for long enough to record. The same can apply to our view of bright planetary disks or lunar features through the telescope eyepiece.

Our perception is also affected by the fact that our brains are wired to pick up movement, and this ability can be translated into a useful observing technique. If a faint object, such as a dim comet, a nebula or a galaxy, is known to lie somewhere within the field of view of an eyepiece but cannot be seen, it can sometimes pop into view when the image is made to wobble slightly – say, by rocking the telescope tube, or by some nifty fingerwork on the instrument's electronic controls.

Adapting to darkness

Two types of light-sensitive cells are found in the retina – cones and rods. Cone cells, found at the center of the retina, give us detailed color vision at the center of our field of view, but are only triggered by bright light. Rod cells, some distance away from the center of the field of view, are triggered in dark conditions. The rods cannot distinguish colors but are far more light-sensitive than the cones. A dim object, such as a diffuse nebula, may be difficult or impossible to see when looked at directly. Using a technique called averted vision, however, observers can shift their view a little way to the side of the actual position of the faint object, so that its light falls on the rod cells. An observer with healthy eyesight may be able to see objects more than two magnitudes fainter using averted vision instead of direct vision.

In very dark conditions, the eye's pupils dilate to their maximum size (around 7.5 mm across in a healthy young adult) allowing the maximum amount of light into the eye. In addition, a light-sensitive pigment called rhodopsin (sometimes called "visual purple") is produced in the retina. Full dark adaptation never happens under urban conditions, since skyglow – caused by stray light from homes, businesses, industry and streetlights – illuminates dust and moisture in the air. Under dark rural skies, stars of the 6th magnitude can be seen by anyone with average eyesight – this is enough for someone with good eyesight to pick out Uranus with the unaided eye. From a light-polluted city, however, often only the brightest objects can be seen. Faint meteors, subtle auroral glows, the zodiacal light and gegenschein are all phenomena denied to the urban astronomer.

A good degree of dark adaptation is required to telescopically observe faint asteroids and to discern subtle detail in the comas and tails of brighter comets. Some observers don a pair of red goggles prior to their observing session, since the eye is less sensitive to red light. Maximum dark adaptation is likely to take a quarter of an hour or so. To optimize dark adaptation, the observer should ensure that there are no local bright sources of light which may be looked at inadvertently. If illumination is required during the observing session – say, to consult a star chart or to sketch an object – a small torch with a red light is essential.

Optical instruments

While a great amount can be seen in the night skies with the naked eye alone, any form of optical aid will allow a closer, more detailed view of objects in the Solar System. Opera glasses, among the simplest optical devices, consist of a pair of basic short-focus refractors, with single objectives and eye lenses. Since their lenses are considerably larger than the eye, they collect more light, reveal fainter objects and resolve more detail than the unaided eye. Despite their optical shortcomings, opera glasses will show the Moon's craters and Jupiter's brightest moons.

The desire to own a quality optical instrument invariably enters the thoughts of most novice Solar System observers. These days, binoculars and telescopes of a tolerably high optical standard are widely available at prices to suit most people's budgets. For example, I frequently carry around an excellent pair of 10 × 25 binoculars which cost (brand new) less than the price of a meal at a fast food restaurant. They are perfect for taking on walks, for general sightseeing and for quick sessions of astronomy, like viewing the Moon or searching for Mercury when it is near the sunset horizon. A small, good-quality budget telescope on a full-length tripod can be obtained for less than the cost of admission to a football game. A sizable Dobsonian reflector or a small but capable computer-controlled instrument can be obtained for less than the price of a season ticket for your favorite football team.

Light-gathering ability and resolving power

The larger the telescope lens, the greater its light-gathering ability. Light from a point source, such as an asteroid, is concentrated into a focused point, so it appears brighter and easier to see through a large telescope than it does through a smaller telescope. Viewed through the eyepiece, the brightness of a point source is proportional to the area of the telescope's objective lens (or mirror). A telescope's light-gathering ability is often expressed in terms of the magnitude of the faintest stars that it will reveal under ideal conditions. The limiting magnitude – the magnitude of the faintest star that can be seen with a given instrument – is the most convenient and meaningful measure of its light-gathering power.

To find the limiting magnitude M visible through a telescope:

$$M = 6.5 - 5\log d + 5\log D$$

TABLE 1 LIMITING MAGNITUDE

Aperture	Limiting magnitude	Example of faint Solar System object visible
Unaided eye	6.5	Uranus
25 mm	9.1	Titan (Saturn's brightest satellite)
60 mm	11.0	Tethys (satellite of Saturn)
100 mm	12.1	Enceladus (satellite of Saturn)
150 mm	13.0	Mimas (satellite of Saturn)
200 mm	13.6	Triton (Neptune's brightest satellite)
250 mm	14.1	Pluto
300 mm	14.5	Titania (Uranus' brightest satellite)

where d is the dark-adapted pupil's diameter (in millimeters), D is the instrument's aperture (in millimeters) and 6.5 is limiting magnitude for stars visible with the unaided eye. d amounts to around 7.5 mm in a healthy young adult in really dark conditions, but will be considerably smaller in light-polluted urban conditions (and so too will the limiting magnitude of the unaided eye). Table 1 is based on an ideal combination of seeing, observer and instrument.

Resolving power, the ability to discern fine detail, increases with telescope aperture. A number of factors determine how high a magnification can be used to deliver acceptably sharp and well-defined images of the Moon and planets. Chief among these is the steadiness of the column of the Earth's atmosphere through which the light from a celestial object passes before reaching the telescope, known to astronomers as "seeing." On nights of good seeing, the maximum useable magnification is around twice the diameter of the aperture (measured in millimeters).

TABLE 2 RESOLVING POWER

Aperture	Max magnification	Resolution (arcsec)	Smallest lunar crater (km)
Unaided eye	1	20"	200
25 mm	50	4.6"	12
60 mm	120	2.0"	6
100 mm	200	1.2"	3.5
150 mm	300	0.8"	2.5
200 mm	400	0.6"	1.8
250 mm	500	0.5"	1.4
300 mm	600	0.4"	1.2

The resolving power for various instruments given in Table 2 represents a practical guide, rather than the theoretical value predicted by the Dawes' Limit equation. To find the Dawes resolving limit R (in arcseconds) for a telescope of aperture D (in millimeters):

$$R = 116/D$$

Seeing

After passing smoothly and uneventfully through space, light from celestial objects is buffeted by the Earth's atmosphere before encountering the astronomer's telescope. Turbulence in the atmosphere affects the quality of the image, a quality astronomers call "seeing." Warm air, released from a landscape cooling down after a warm day, or from houses and industry, interacts with the colder night air, producing a turbulent mix of different densities and refractive properties.

In addition, the flow of the air can be disrupted by hills and buildings, forming eddies which further mix and degrade the seeing. Stars nearer the horizon, whose light passes through a greater amount of atmosphere, are most prone to scintillation. The Moon and planets, however, being extended objects rather than point sources, do not usually appear to twinkle, but that's not to say that they are immune from the effects of turbulence. Through the eyepiece, atmospheric turbulence can play havoc with lunar and planetary observation, sometimes being so bad as to render observations futile.

A simple scale of seeing, devised by the great planetary observer Eugène Antoniadi (1870–1944), can be used to describe the quality of the atmosphere at the time of each observation. The Antoniadi Scale is extensively used by UK observers:

AI Perfect, image exceptionally steady.
AII Good, generally steady image with occasional undulations.
AIII Moderate, frequent undulations with some moments of calm.
AIV Poor, undulations constant.
AV Appalling, hardly worth observing.

On the basis of observations made through a 125 mm refractor, William H. Pickering (1858–1938) of Harvard Observatory devised a more scientifically based scale of seeing. The steadiness of the atmosphere was gauged by the visible detail and extent of the Airy disk (the small circle of light into which the light from a point source, such as a star, is concentrated) and diffraction patterns (artifacts surrounding a star's image caused by the instrument's optics). Users of the scale rarely go to the lengths of actually assessing the visible diffraction patterns, since regular observers are quite capable of accurately judging the quality of an image and assigning its position on a scale of 1 to 10. In the scale listed below, 1–2 is very poor, 3–4 is poor, 5 is moderate, 6–7 is good, 7–8 very good, and 8–10 is excellent.

P1 Star image is usually about twice the diameter of the third diffraction ring if the ring could be seen; star image 13″ in diameter.
P2 Image occasionally twice the diameter of the third ring (13″).
P3 Image about the same diameter as the third ring (6.7″) and brighter at the center.
P4 The central Airy diffraction disk often visible; arcs of diffraction rings sometimes seen on brighter stars.
P5 Airy disk always visible; arcs frequently seen on brighter stars.
P6 Airy disk always visible; short arcs constantly seen.

P7 Disk sometimes sharply defined; diffraction rings seen as long arcs or complete circles.
P8 Disk always has sharply defined rings seen as long arcs or complete circles, but always in motion.
P9 The inner diffraction ring is stationary. Outer rings momentarily stationary.
P10 The complete diffraction pattern is stationary.

Binoculars

Binoculars are a frequently underestimated tool in the Solar System observer's optical armory. By using prisms to fold the light path, their length is reduced to manageable dimensions, and they usually have a right-way-up view so that they can also be used for terrestrial viewing.

Porro prism binoculars are the most widely available type. Traditional Porro prism designs have the familiar W-shape, which is the result of the light path being folded inward from widely spaced objective lenses toward the observer's eyes. Some small Porro prism binoculars have a U-shaped shell and use inverted Porro prisms, which means that their objective lenses can be positioned closely adjacent to each other – in some cases closer together than the distance between the observer's eyes. The difference between the two types of Porro prism binoculars has no implications for astronomy, but the stereoscopic effect when viewing nearby terrestrial objects is better with the more widely spaced objectives of the more conventional design.

Roof prism binoculars – so-called because of the particular shape of their prisms – are compact and lightweight. Most have a straight-through shape, resembling two small and ordinary refracting telescopes side by side. Inside the straight barrels of these binoculars, roof prisms fold the light path five times, producing a fairly long focal length instrument in a deceptively short body. These instruments are very handy to carry around in the coat pocket or handbag to snatch low-power views of the skies. Capable of

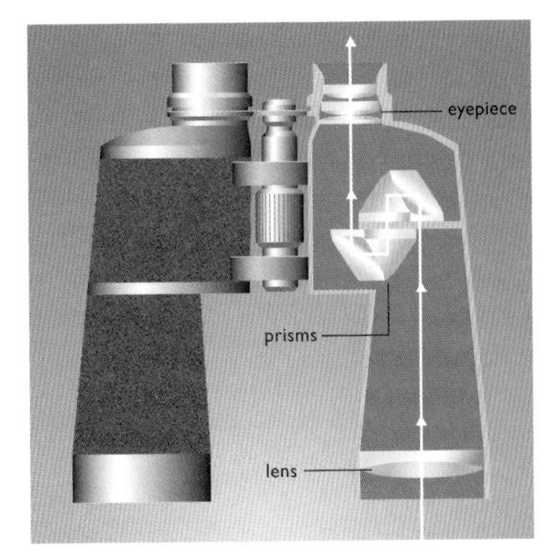

▶ *The optical configuration of the most common form of binocular. It uses a combination of accurately aligned lenses and Porro prisms.*

revealing literally hundreds of the larger lunar craters, a pair of 10 × 25 binoculars is ideal for quick sessions of lunar observing, viewing appulses between the Moon, stars and planets.

Binoculars are by far the best starter instrument for an amateur astronomer. They are labeled with two numbers, which indicate their magnification and aperture. For example, a pair of 7 × 50 binoculars (the best specification for general stargazing) delivers a magnification of 7 and has lenses of 50 mm diameter. Highly portable, such binoculars can be used in corners of the backyard inaccessible to telescopes, and can be carried to dark sky sites with ease.

By holding a pair of binoculars up to the light and looking at the eye lenses from a distance, a small circle of light can be seen. The diameter of this circle is called the "exit pupil," and it can be calculated by dividing the size of the binoculars' objectives by their magnification. A pair of 12 × 50s has an exit pupil of about 4 mm, while a pair of 7 × 50s delivers an exit pupil of around 7 mm – about the size of a fully dark-adapted pupil. The larger the exit pupil, the better suited the binoculars to general astronomy since they deliver brighter views. If the exit pupil is larger than the eye's pupil, however, the binoculars will not deliver all of their light to the retina, and they will underperform.

Binoculars can be used for projecting the Sun's image onto a shielded white card (see chapter 15) – remembering of course to bear safety in mind and to keep the lens cap on one half of the binoculars. Experienced observers have constructed safe filters to cover the objective lenses of binoculars when viewing the Sun, and commercial solar binoculars are also available. Using these in public, however, might give entirely the wrong impression to the young and/or inexperienced – leading to the disastrous misconception that it's perfectly safe to view the Sun directly.

Viewed at a low magnification through binoculars, the Moon appears almost three-dimensional, especially during total eclipses or when there is a degree of earthshine. A small pair of binoculars will reveal the crescent phase of Venus, the disk of Jupiter and the Galilean moons, Saturn and its largest satellite Titan. Many minor planets can be followed through binoculars – some 25 asteroids brighten to magnitude 10.2 or above at opposition, and are therefore detectable through a pair of 50 mm binoculars. With their wide fields of view, binoculars show comets to excellent effect, with usually several visible each year through binoculars. Although 50 mm binoculars have the same light-gathering ability as a 50 mm refractor of equivalent focal length, using two eyes instead of one allows more detail to be discerned, and it makes for a more pleasant and relaxing experience.

For general stargazing, a pair of 7 × 50 or 10 × 50 binoculars is ideal. They deliver a wide field of view and a bright image, and have a magnification low enough to be able to smoothly and comfortably scan the skies for short periods. A field of view about 7° across is delivered by a pair of 7 × 50s – that's an area of sky nearly 200 times larger than the full Moon.

When observing Solar System objects, binoculars need to be held firmly, with some kind of support to keep them steady and fixed on the object under scrutiny. It is virtually impossible to enjoy a steady view using hand-held binoculars of magnifications greater than 7×. Physical comfort is also essential. An observer using binoculars attached to an ordinary photographic tripod to view objects at a high angle above the horizon needs to crane his or her neck quite considerably. In addition, highly tilted binoculars on such a tripod may become somewhat out of balance, causing more problems. Binoculars with angled eyepieces (45° or 90°) help a great deal, since the observer's head is tilted forward rather than backward. Another solution is to use a parallelogram mount – a device that allows the observer to access a large area of the sky without greatly altering position.

Giant binoculars – those with objectives larger than 70 mm – require a very firm and steady hold, but they deliver highly pleasing views of the Moon and planets. Through a pair of 20 × 80 binoculars, lots of lunar detail can be observed, Venus' phases can be followed, Jupiter's disk is clear (along with traces of its cloud belts), and Saturn's rings can be discerned. It may be tempting to purchase a pair of giant zoom binoculars – say, a pair of 20–100 × 70 binoculars – but they are best avoided unless of a premium brand, since any slight defects in quality become exaggerated when higher magnifications are used.

▲ A comparison between the Moon's apparent size when viewed through 7 × 30 binoculars (left) and a hefty pair of 25 × 100s.

Telescopes

Many amateur astronomers begin with a small, unsophisticated telescope and upgrade to larger instruments as their budget allows. For a beginner, small telescopes have a number of advantages, being lightweight, portable, and to some extent expendable. Small telescopes are often provided with a selection of two or three eyepieces. The maximum useable magnification for any telescope is around twice the objective's size in millimeters, for example, 100× for a 50 mm telescope; higher magnifications than this will not produce a better, clearer image. While the optical quality of a small budget telescope's objective lens or mirror may be good, the eyepieces provided may not be good enough to get the best performance from the instrument. Star diagonals may have such poor-quality mirrors that they are unusable, and the Barlow lens – an accessory that doubles or triples the magnification delivered by any eyepiece – is not likely to improve the view. By using good-quality eyepieces which deliver low-to-moderate magnifications, the instrument will earn its keep, and the observer will enjoy some lovely views of the Solar System. It is worth obtaining one or two good-quality 1.25-inch barrel diameter Plössl eyepieces, along with a 0.96–1.25-inch barrel diameter eyepiece adapter if required.

Refractors

Refracting telescopes consist of a tube with an objective lens at one end, which collects and focuses light, and an eyepiece at the other, which magnifies the focused image. The focal length of a refractor is expressed as a multiple of the lens diameter – a 100 mm diameter objective of *f*/10 will have a focal length of 1000 mm. The most basic type of refractor has a single objective lens and a single eyepiece lens. Such refractors require very long focal lengths in order to combat the effects of false color, known as chromatic aberration, which is caused by the different degrees to which the constituent colors of light are refracted by the objective lens. Through this type of refractor, bright objects like the Moon and Venus are surrounded by fringes of colored light.

Most of the effects of chromatic aberration can be overcome by using an achromatic objective lens, which consists of two specially shaped lenses sandwiched together. Achromatic objectives focus the different wavelengths of visible light near to a single point, but do not entirely eliminate false color: residual chromatic aberration is particularly noticeable in budget-value short-focus refractors of *f*/8 and below. Although the edges of bright objects like the Moon and planets display a violet tinge, it is aesthetically unobtrusive to most observers and can be almost eliminated with a minus-violet filter fitted into the eyepiece.

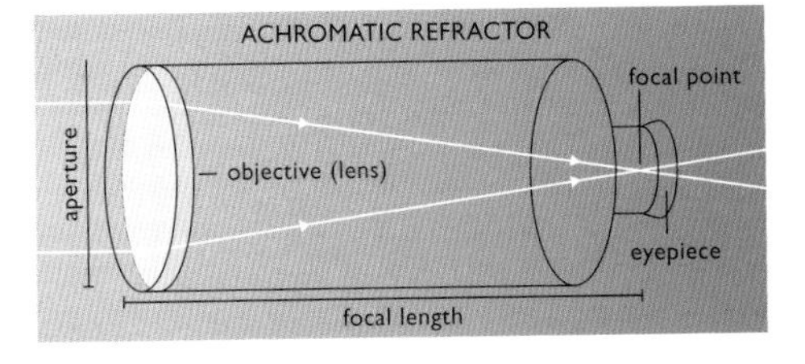

▶ *Refractors gather and focus light using a large objective lens.*

It is also possible to correct the color aberrations in refractors by using a special correcting lens which fits into the focuser; although this option can be expensive, it is cheaper than buying a fully fledged apochromatic instrument.

Apochromats deliver images free from chromatic aberration by using special glass in their two- or three-element objective lenses; these focus all wavelengths of visible light into as near a single point as possible. Capable of delivering near-perfect, high-contrast images, apochromats are among the most desirable telescopes for lunar and planetary observation. On average, however, they are ten times costlier than achromats of a similar-size.

Reflectors

By using mirrors to collect and focus light, reflecting telescopes produce images that are free from the effects of chromatic aberration. Newtonian reflectors are the most commonly used type. They have a concave paraboloidal primary mirror at the base of the telescope tube, which reflects light to a small flat secondary mirror near the top of the tube, which in turn reflects the light out of the side of the tube and into the eyepiece. This secondary mirror obstructs a small portion of the light path entering the telescope, which means that a Newtonian of a certain aperture has a smaller light-gathering surface than a refractor with an objective lens of equivalent diameter. In addition, the secondary mirrors and their supporting structures intrude into the light path, causing some distortion of the image – spikes can be seen around stars caused by scattering of light. This phenomenon, known

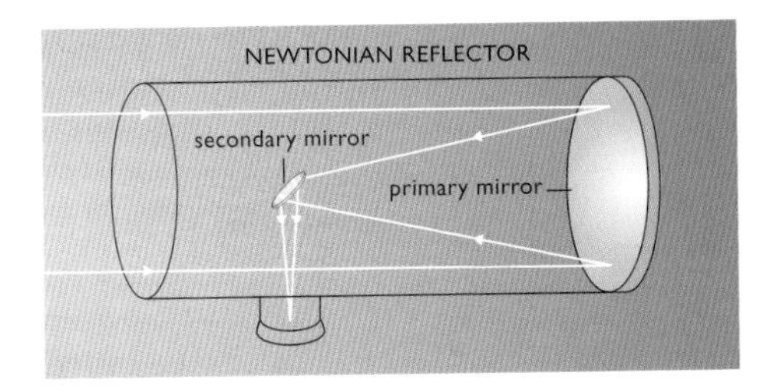

▶ *Reflectors use a large concave primary mirror to collect and focus light. In Newtonian reflectors, a small flat secondary mirror diverts the light into an eyepiece at the side of the tube.*

as diffraction, means that the image loses a certain amount of contrast – a property most valuable in being able to discern subtle lunar and planetary detail.

The effects of diffraction in Newtonian reflectors become less severe with longer focal lengths, since they have smaller secondary mirrors. A 150 mm, *f*/8 Newtonian reflector, one of the most commonly produced specifications for a Newtonian, is capable of giving pleasing all-round celestial views, and can be used at high magnifications for Solar System studies. For detailed high-magnification Solar System observation, even longer focal lengths will perform better. A well-collimated long focal length Newtonian reflector – say one of *f*/10 or greater, with high-grade optics – is capable of delivering views that are of apochromatic quality. The chief disadvantage of long focal length Newtonians is the length of their tube – a 250 mm, *f*/10 will have a focal length of 2.5 m. Such an instrument will require a substantial, sturdy tube and mount, whereas a 250 mm, *f*/5 Newtonian is quite capable of being used as a portable instrument on a Dobsonian mount.

▲ *An 8-inch Newtonian reflector on an equatorial mount. Note the curved vanes holding the secondary mirror – these eliminate diffraction spikes around bright stellar objects.*

Eminently suited to high-powered lunar and planetary observation, Cassegrain reflectors have contributed greatly to Solar System studies. Since they have a reputation for being difficult instruments to make, maintain, adjust and collimate, their use has never been very widespread among amateur astronomers. They have a concave primary mirror, which reflects light onto a convex central secondary mirror, which in turn reflects the light back down the tube, through a hole in the primary and into the eyepiece. With their twice-folded light paths, Cassegrains pack a lot of focal length into their tube – usually a skeleton tube, to cut the weight of the instrument and to minimize the effects of tube currents. Cassegrains are prone to the twin optical aberrations of astigmatism and field curvature, and for these reasons they are usually found in very high focal lengths, from *f*/15 to *f*/25.

▲ *View down the tube of a Newtonian reflector. The primary mirror shows the reflection of the secondary. Note the single curved vane holding the secondary mirror, an arrangement that eliminates diffraction spikes in the image.*

Catadioptrics

Catadioptric telescopes use both mirrors and lenses to collect and focus light. Schmidt–Cassegrain telescopes (SCTs) are the most popular catadioptric used by modern amateur astronomers and Solar System observers. A hybrid of the Schmidt camera and the Cassegrain telescope design, SCTs are closed systems consisting of a perforated primary mirror and a secondary mirror fixed to the inside of a whole aperture lens. Known as a correcting plate, the lens appears thin and flat, but is very minutely figured in order that the light passing through it is slightly bent so that the rays are reflected to a focus by the concave spherical primary mirror. Light is reflected from the convex secondary through the central perforation in the primary mirror, into the eyepiece. In common with SCTs, Schmidt–Newtonian telescopes (SNTs) have a full-aperture frontal corrector plate, and their secondary mirror reflects light out of the side of the tube into the eyepiece.

With a relatively large central obstruction, diffraction within SCT and SNT systems has the potential to compromise very slightly the quality of the image when viewing detail on the Moon and planets – a point often overemphasized by zealous telescope purists. In truth, a well-collimated SCT or SNT with good optics will produce really excellent views of Solar System objects.

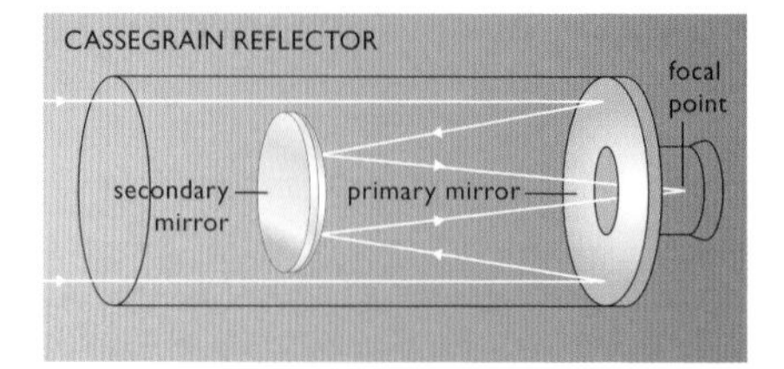

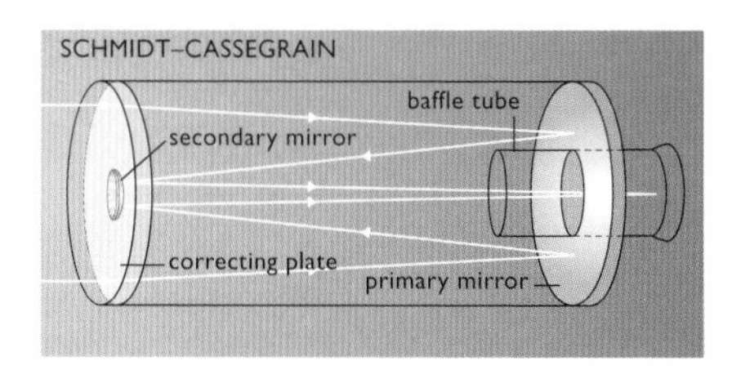

◀ *Both traditional Cassegrain and Schmidt–Cassegrain (SCT) telescope designs use a small secondary mirror to intercept and focus light through a hole in the primary mirror and into an eyepiece behind it.*

SCTs can easily accommodate a wide range of useful accessories, such as filter wheels, cameras, webcams, digicams, camcorders and CCD cameras. When viewing objects directly through the eyepiece, an SCT produces a "traditional" inverted astronomical view. For ease of use, SCT users often use a star diagonal, which gives a mirror image of the already inverted image. Observers might find it a little tricky at first to find their bearings – deciding whether the Moon is about to occult a star, or already has, or whether a spot on Jupiter is about to

▲ *Maksutov–Newtonian telescopes (MNTs), such as this 150 mm example belonging to Jamie Cooper, offer crisp, contrasty views of the Moon and planets.*

transit the central meridian, or already has, can require a moment or two to figure out.

Maksutov–Cassegrain telescopes (MCTs) use a spherical primary mirror and a deeply curved spherical meniscus lens at the front of the tube to correct for spherical aberration. The secondary mirror itself is a small spot that has been aluminized directly onto the interior of the meniscus lens. In their general appearance, MCTs resemble SCTs in many ways, but with their longer focal lengths, their performance on Solar System objects is generally better. They offer near-apochromatic quality, high-resolution, high-contrast views of the Moon and planets without any appreciable chromatic aberration.

Catadioptrics are sealed systems, and the only optical surface directly exposed to the outside world is the exterior of the corrector plate or meniscus. Great care must be taken when getting rid of any debris that accumulates on this component, as it is often coated with contrast-enhancing material. SCTs can come out of collimation over time, but this can usually be corrected by means of three small collimation screws on the exterior housing of the secondary holder. Being closed systems, catadioptrics require a certain cool-down time to equalize their internal temperature and that of their optics with the outside air – this process can be sped up by means of an internal fan in some larger catadioptrics. Treated with care, catadioptrics require much less maintenance than reflectors, and their mirrors can last for decades without requiring re-aluminizing. Perhaps the biggest downside of these versatile instruments is that their corrector plates or menisci require protection from dew, which is liable to accumulate very fast given certain conditions, putting an end temporarily to any observing session. A dew cap or electrical dew zapper (a very mild heating element which wraps around the edge of the corrector plate housing) is an essential requirement if you want to follow Solar System objects through all the seasons of the year.

Eyepieces

New observers often overlook the fact that the eyepiece is as important to the performance of a telescope as its objective lens or primary mirror. A poor-quality eyepiece will deliver a substandard image, frustrating the observer, and possibly acting as a deterrent to further observational astronomy. Many budget instruments are provided with inadequate eyepieces, which is a great pity, since the telescope optics themselves may be of excellent quality.

Eyepieces serve to magnify the image. To calculate the magnification provided by a given eyepiece, divide the focal length of the telescope by the focal length of the eyepiece. For example, a 9 mm eyepiece will give a magnification of 133× in a telescope with a focal length of 1200 mm (1200/9 = 133). Magnification is important to Solar System observers, but should not be the sole factor influencing the choice of eyepiece. The apparent field of view must also be considered. This is the width of the image visible through an eyepiece, expressed in degrees, when the eye is at the optimum distance from it. A larger apparent field of view tends to enhance the observing experience; small fields of 40° or less give the impression of peering up the shaft of a deep well at the night sky above. Apparent fields of view larger than 70° are so wide that they are often described as a "spacewalk" experience – so wide that one's eye cannot take in the whole field at once – but such an experience does not come cheap.

Three good-quality eyepieces giving different magnifications should be enough to satisfy the basic needs of the Solar System observer:

1. A low-power eyepiece with a fairly wide actual field of view. It should be wide enough to accommodate the whole solar or lunar disk (0.5°), or to fit in a couple of planets during a close appulse.
2. An eyepiece of medium power (up to 100×) for more detailed observation.
3. A high-power eyepiece. The highest reasonable magnification to use on nights of average seeing can be found by doubling a telescope's aperture in millimeters, for example, 300× for a 150 mm reflector. This eyepiece can be used for observing fine solar, lunar and planetary detail.

Of the three eyepiece barrel diameters generally available – 0.965-inch, 1.25-inch and 2-inch – the most common are 1.25-inch. An increasing number of focusers are fashioned with an adaptor to accept both 1.25-inch and 2-inch diameter eyepieces. Two-inch eyepieces are usually of the low-magnification, wide-angle variety.

Basic eyepieces

Huygenian, Ramsden and Kellner type eyepieces are very old designs delivering small apparent fields of view. They are usually provided with small budget instruments, but do not meet the demands of the modern Solar System observer.

Good eyepieces

Orthoscopic eyepieces produce a flat, aberration-free field, and deliver very good high-contrast views of the Moon and planets. Their apparent field of view is around 50° and they have a reasonable degree of eye relief.

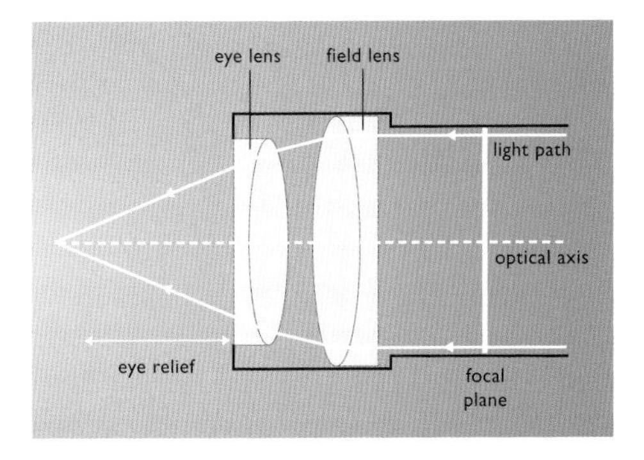

► *A simplified cross-section through a Plössl eyepiece – one of the most widespread designs in use.*

Plössl eyepieces are probably the most popular type among Solar System observers. They have minimal internal reflections and good color correction. The apparent field of view, around 50°, is flat and sharp even up to the edges. Their main disadvantage is that the shorter focal length eyepieces have poor eye relief, which means that the eye has to be very close to the lens to take in the whole field of view of a high-magnification scene. Longer focal length Plössls give acceptable eye relief, and there are special long eye-relief versions with large eye lenses for wearers of spectacles.

Monocentric eyepieces are an old design, developed in the 19th century by Carl Zeiss, but they offer probably the best performance for high-magnification lunar and planetary observing. Although they deliver only a tunnel-like 30° field of view, Monocentric optics are virtually scatter-free, producing high-contrast images, the quality of which cannot be excelled by any other eyepiece.

Erfle eyepieces have a somewhat wider apparent field of view, and give a low-magnification field with good color correction. They perform at their best when used with long focal length telescopes. Definition at the edge of their 70°-wide apparent field of view is liable to be on the soft side, however, and ghost images of bright lunar and planetary subjects may be visible due to internal reflections.

Wide-angle eyepieces

Introduced in recent decades, modern wide-field and ultra-wide-field eyepieces like the Meade Ultrawides, Celestron Axioms, Vixen Lanthanum Superwides and Tele Vue Radians, Panoptics and Naglers deliver excellently corrected images with very large apparent fields of view. Large fields are useful on an undriven telescope because the subject stays in the field for longer and the telescope's aim does not require adjusting as often. Naglers, with their 80° plus apparent fields, are splendid eyepieces which deliver breathtaking wide-angle views. The

TABLE 3 ACTUAL AND APPARENT FIELDS FOR DIFFERENT EYEPIECES		
Magnification	Apparent field	Actual field
25	50° / 80°	2° / 3.2°
50	50° / 80°	1° / 1.6°
75	50° / 80°	41′ / 1.1°
100	50° / 80°	30′ / 48′
150	50° / 80°	19′ / 32′
200	50° / 80°	15′ / 24′
300	50° / 80°	10′ / 16′

TABLE 4 VARIOUS EYEPIECES USED ON A 200 MM *F*/10 SCT			
Eyepiece	Magnification	Apparent field	Actual field
Meade Series 4000 (56 mm)	36	52°	1.4°
Celestron Ultima (42 mm)	48	36°	45′
Meade Series 4000 SWA (32 mm)	63	67°	1.1°
TeleVue Nagler (31 mm)	64	82°	1.3°
Meade Series 3000 (25 mm)	80	50°	38′
Celestron Axiom (19 mm)	105	70°	40′
TeleVue Nagler (17 mm)	118	82°	42′
TMB Monocentric (14 mm)	143	30°	13′
TeleVue Nagler (7 mm)	286	82°	17′
Celestron Ultima (5 mm)	400	50°	5′
TMB Monocentric (4 mm)	500	30°	4′

TABLE 5 SOME SOLAR SYSTEM DIMENSIONS	
Average diameter of solar corona seen at eclipse	1.5°
Diameter of Moon	29.5′–33.5′
Appulse of Jupiter and Venus (give date example)	
Maximum diameter of Callisto's orbit around Jupiter	20′
Maximum distance of Titan from Saturn	3.5′
Maximum diameter of Venus	1.1′
Maximum diameter of Jupiter	50″
Maximum width of Saturn's rings	45″

longer focal length Naglers are big and very heavy – the 13 mm Nagler is 135 mm high and weighs a hefty 0.7 kg (25 oz) – so switching between Naglers and regular eyepieces will almost certainly require the telescope to be rebalanced each time.

Zoom eyepieces are no longer just a novelty item. A number of quality zooms are on the market, allowing the observer to change between focal lengths of, say, 8 to 24 mm at the twist of the eyepiece barrel. Their apparent field of view narrows from a generous 60° at 8 mm to a somewhat miserly 40° at 24 mm, but if you can accept this restriction, a single zoom eyepiece can replace a drawer full of regular Plössl eyepieces. While most zoom eyepieces require a little refocusing after changing their focal length, the Nagler zoom remains in focus throughout its 3 to 6 mm range.

Zoom eyepieces go well with the binocular viewer, a device that splits the single beam of light from the objective and diverts it into two identical eyepieces. Most binocular viewers will work only on a refractor or an SCT; they require a long light path, which means that it may not be possible to focus one using a Newtonian reflector. Binocular viewers can only be used with two identical eyepieces, and these ought to be of a focal length shorter than about 25 mm in order to prevent image degradation around the edge of the field of view. A binocular viewer adds a sense of perspective to observations of Solar System objects, producing an almost three-dimensional quality. It has been demonstrated that viewing with both eyes enables much finer detail to be discerned.

Barlow lenses and focal reducers

An extra lens can be placed in the light path before it enters the eyepiece in order to extend or shorten the effective focal length of the telescope, altering the magnification delivered by any given eyepiece.

A negative lens called a Barlow lens effectively increases a telescope's focal length – usually doubling or tripling it – increasing the magnification delivered by an eyepiece. Barlow lenses are commonly used for lunar and planetary observation – indeed, it is often better to use them for high-magnification work rather than switching to a shorter focal length eyepiece which may have poorer eye relief. Barlow lenses are often used in combination with webcams for lunar and planetary imaging at prime focus, since the image requires enlarging as much as possible on the low-resolution matrix of the CCD chip in the webcam.

A focal reducer/corrector is a positive lens frequently employed on SCTs. The most common type reduces the focal length of a typical *f*/10 SCT to *f*/6.3, which reduces the magnification of any given eyepiece by 37%, producing an actual field of view some 160% wider and 2.5 times greater in area. A reducer/corrector is ideal for taking in wide fields of view, to see the entire lunar or solar disk, to view eclipses, and to hunt and observe comets. Reducers/correctors also flatten the field of view to the edge, making them a better option for use with a given eyepiece than a lower magnification eyepiece, which may show a degree of edge distortion when used without the reducer/corrector. Shorter astrophotography exposure times are possible using reducer/correctors, and they deliver a larger field on the CCD chip.

Telescope mounts

Telescope mounts require a high degree of rigidity and stability. Altazimuth mounts allow the telescope to be moved up and down and swung from side to side. The most popular type of altazimuth mount is the Dobsonian. Simple and easy to use, Dobsonians consist of a box with an azimuth bearing that rests on the ground; this carries another box (and the telescope tube) with an altitude bearing at its center of balance. Low-friction materials like polythene, Teflon and Formica are used for the load-bearing surfaces, which allows easy fingertip control. Lightweight structural materials such as plywood make Dobsonian-mounted instruments very portable.

The Earth's rotation causes every celestial object to make a complete circuit around the celestial pole every 24 hours. An object located near the celestial equator will appear to move westward at the rate of one quarter of a degree per minute. This may not seem much – indeed, the motion cannot be detected with the unaided eye – but once the sky is magnified through the telescope eyepiece, it becomes apparent. A small telescope on an undriven altazimuth mount can be used to observe the Moon and planets using lower powers only, up to a maximum of 100×; higher magnifications will not permit the frequent adjustments necessary to keep an object in the field of view long enough for it to be properly observed. An eyepiece with a 50° apparent field giving a magnification of 100× will take in an actual field of view about half a degree wide, which is enough to fit in the whole Moon. Without a drive, an object – such as a lunar crater or Jupiter – will appear to drift from the center of the field to the edge in less than a minute. Making the light adjustments needed to keep an object centered within the field of view can be a difficult task using telescopes mounted on photographic tripods or simple altazimuth mounts with clamp-friction bearings.

Equatorial mounts allow the telescope to follow celestial objects as the Earth rotates. A properly set-up equatorial mount, with its polar axis aligned with the celestial pole, allows the observer to keep a celestial object centered in the eyepiece with ease. Even if the mount has no motorized drive, a light push on the tube or a slight twist of the slow-motion knob on the RA axis will keep the object in view – a far easier task than the pushing and pulling required with an altazimuth mount. A motor drive makes for the most hassle-free viewing experience.

Two types of equatorial mount are in widespread use. German equatorial mounts are sturdy and flexible, and are often used to mount refractors and reflectors. The entire sky above the horizon, including the celestial pole, is accessible with a German equatorial mounted telescope. SCTs are usually mounted between the arms of heavy-duty fork mounts. Many SCTs are capable of accurate computer-controlled

► *Dobsonian-mounted telescopes provide a smooth push-pull altazimuth motion, enabling the entire sky to be accessed with ease. Dobsonians accommodating Newtonian reflectors are both lightweight and portable.*

► *Many small budget refractor telescopes are provided with simple altazimuth mounts. Telescopes on such basic mounts are capable of being used only at low magnifications, since they require manual operation. A number of more sophisticated computerized Go-To SCT telescopes are also mounted in the altazimuth style, providing accurate location and automatic tracking of celestial objects.*

► *Once the most common form of telescope mounting, the German equatorial has two axes of rotation – one axis is aligned with the celestial pole, while the other axis is perpendicular to it. Once centered in the field of view, a celestial object can be tracked with ease. A counterbalance to the telescope (be it a refractor, reflector or SCT) is required, which decreases the mount's portability.*

► *Fork mountings are most suited to large reflectors or SCTs. The telescope is capable of moving in Declination between two hefty arms, while the fork's axis of rotation is tilted toward the celestial pole and can move in Right Ascension. Large computerized SCTs on polar-aligned fork mounts are popular with Solar System observers and CCD imagers alike.*

equatorial tracking in altazimuth mode, both axes being adjusted continually. For greater tracking accuracy, the whole fork mount can be tilted to point to the pole, requiring only one axis to be driven to keep an object centered in the field. Fork-mounted instruments are often denied a view of a small region near the celestial pole, since any large eyepieces or accessories, such as binocular viewers or CCD cameras, will not permit the telescope to move between the fork and the very base of the mount.

Computers and astronomical programs

A number of comprehensive astronomical programs displaying detailed graphical views of the sky and the positions of Solar System objects are available for the personal computer – some very good ones are available as freeware for download from the Internet. They can be consulted to discover the past, present and future positions of Solar System objects, along with all the data needed about each object in question – its magnitude, apparent angular diameter, illumination, and so forth. As planning tools and observational aids, planetarium programs are invaluable. The best programs allow magnified views of each object, showing the phase and shadows of a planet and the exact position of its satellites (if it has any) and their phenomena (such as eclipses). Instant program updates for the latest asteroid and cometary positions are available via the Internet for premium planetarium programs.

With the advent of portable laptop computers and pocket-sized personal digital assistants (PDAs), there is no need for an observer to be without the necessary information in the field. Laptops and PDAs are capable of guiding telescopes, enabling them to slew to a vast range of celestial objects. Laptops can provide instant processing facilities for astronomical CCD imagery.

Computer power and new technology has the potential to enhance the Solar System observer's viewing experience. Still, nothing can compare with the live view of the Moon or a bright planet through the eyepiece. It's not unusual for a first-time observer to let out an involuntary gasp of astonishment once Saturn enters the eyepiece of even a small telescope, but it's doubtful that a simulated image of the ringed planet on a PC monitor could ever produce the same feeling of awe!

Recording the Solar System

Observational drawing

With the growth in popularity of digital imaging has been a decline in the number of Solar System observers making observational drawings of their views through the eyepiece. It is, however, quite wrong to imagine that observational drawing is an arcane activity of no practical use today. On the contrary, the need for visual observers to make accurate sketches and notes at the eyepiece is greater than ever, for there is no other way of making a true record of what can actually be seen at the eyepiece. Without visual observations, there's nothing to challenge the accuracy of what is recorded on an electronic image, which might, through a glitch in the hardware or by misprocessing, display erroneous data that is taken for a real feature.

Much fine planetary detail visible through the eyepiece is of such a subtle nature that it takes a trained observer to see it, let alone record it with accuracy. CCD images are invariably processed in order to bring out this faint detail so that it is glaringly obvious to the viewer, thus creating a misrepresentation of the actual scene. In turn, the proliferation of such sharp and highly detailed images of the Moon and planets can produce apathy on the part of visual observers – why bother to go to the trouble of spending an hour or two observing and drawing astronomical objects when the CCD can apparently capture it all with great accuracy in a fraction of a second? Why observe at all, when images captured by a CCD will bring the scene live to one's computer screen indoors? These are great questions, but they are asked only by those who don't get a thrill from seeing the heavens for real.

Observational drawing is an immensely useful and rewarding activity which improves every single aspect of one's observing skills. The observer's ability to discern fine detail on the Moon and planets constantly improves with time spent at the eyepiece. By concentrating on

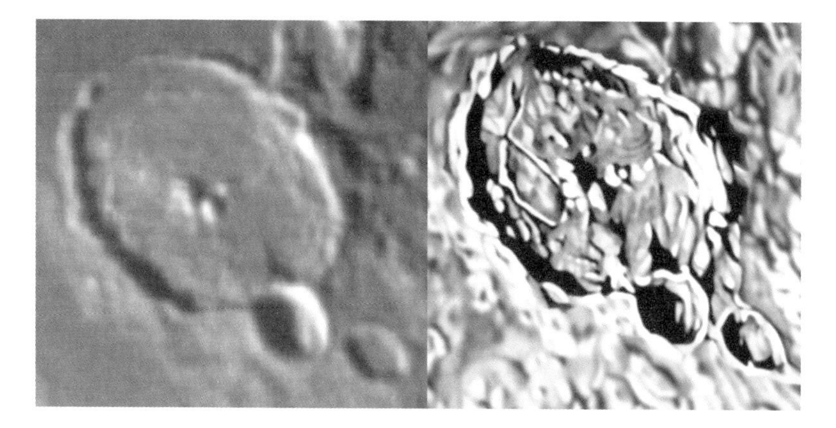

▲ *Comparison between a CCD image (left) and a drawing of the lunar crater Gassendi, both made on April 23, 2002, by the author, using a 150 mm reflector.*

drawing an object, rather than peering at it in a cursory fashion, the observer learns to attend to detail instead of allowing the eye to wander around the more obvious features.

Drawing the Moon's features to the observer's satisfaction is one of the most challenging tasks an amateur astronomer can undertake. At first, it may seem daunting, since there's so much detail visible through the eyepiece of even a small telescope. Armed with a good lunar map, however, the apparent confusion of the Moon's landscape becomes increasingly familiar. It is important to have a degree of confidence in your own drawing abilities. Points are not given for artistic flair or aesthetic appeal – observational honesty and accuracy counts above all. Drawing the planets requires a great deal less pencil work than lunar sketching. Preprepared observing blanks are used: standard 50 mm diameter circles are used for Mercury, Venus and Mars; a larger ellipse is used for Jupiter; and Saturn requires a specially drawn template showing a suitable tilt of the planet's rings. The satellites of Jupiter and Saturn, along with any background stars, are best plotted on a separate drawing on the same sheet of paper, with the planet drawn to a smaller scale. Small comets can be centered in the field of view, and the surrounding starfield plotted; using a higher magnification eyepiece, detail in the coma surrounding the nucleus can be sketched. Note that during the course of an observing session, the comet's movement against the background may be noticeable.

After securing an observational drawing at the eyepiece, a neat copy is best made as soon as possible after the observing session, while the visual information is still fresh in your mind. Retain the original drawing, since it can be used as the basis for any subsequent copies that need to be made. Photocopies of observations are not good enough to submit to the observing sections of astronomical societies or for publication in magazines, but most will accept a good laser print or digitally scanned drawing. Some magazines prefer artwork to be submitted on disk or by email.

Conventional photography

Conventional cameras excel at capturing auroral displays, meteor showers, wide-angle views of comets, and lunar and planetary appulses, providing that the exposure time can be determined by the user. In addition to making observational drawings of the projected Sun, solar observers often photograph the projected image. Care must be taken to position the camera as close as possible to the light path exiting the telescope eyepiece (without actually getting in its way) since the round solar disk appears distorted into an oval shape when viewed at anything other than a square-on angle.

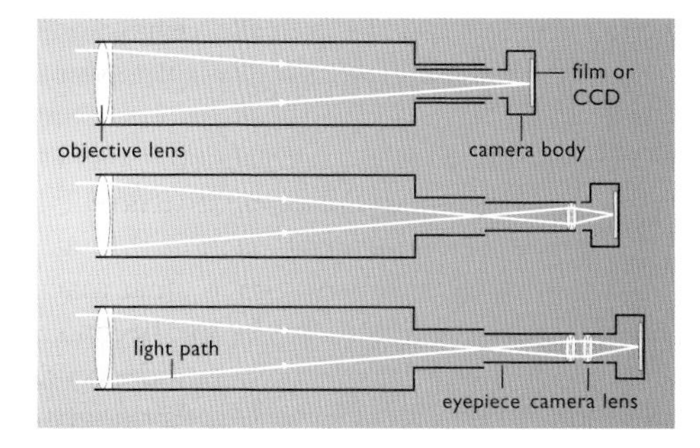

◀ *Imaging is possible with different combinations of camera and telescope: prime focus (top), eyepiece projection (middle) and afocal (bottom).*

Afocal photography

Opportunities for imaging Solar System objects with an ordinary camera held close to the telescope eyepiece are limited. The technique, called afocal photography, requires the camera to be held steadily near the eyepiece while the shutter is released. Pleasing images of the Moon can be obtained in this way, using a regular camera – even a single-shot disposable camera – through a low-magnification eyepiece of an undriven telescope. Such a hit-and-miss approach is liable to produce a fair number of blurred or ill-framed exposures, however, so it can be rather a costly exercise. Although the major planets may appear large enough and bright enough through the telescope eyepiece to make good subjects for afocal imaging, the results often disappoint. Conventional planetary photography needs a rock-steady driven mount, since any slight movement of a planet during an exposure will cause blurring. A 35 mm compact camera or a digital camera can be attached to a mount which is anchored firmly to the telescope, allowing the distance between the camera lens and the eyepiece to be adjusted.

Photographs taken afocally are often surrounded by a dark circular border, a vignetting effect caused by the restricted field of view through the eyepiece. The degree of vignetting depends on the size of the camera's fixed lens in relation to the eyepiece lens, the type of eyepiece used and the distance of the camera lens from the eyepiece. With some experimentation, it should be possible to minimize the vignetting the effects.

SLR photography

Until the advent of digital cameras, the 35 mm SLR (single lens reflex) camera was the most popular piece of imaging kit among astrophotographers. The SLR's optical configuration gives a direct view of an object as it will appear on the image on film – an advantage not shared by ordinary compact 35 mm cameras. Since the SLR lens can be interchanged, the camera can be used for capturing a variety of images, from wide-angle short-exposure views of the Moon and planets, to wide-field cometary studies with telephoto lenses, to close-up images of Solar System objects. In the last-mentioned field, however, SLRs have been surpassed by digital cameras: CCD chips are far more sensitive to light than is photographic film, and hence require shorter exposures, minimizing blurring.

If the camera body is attached to the telescope using a T-adapter, the telescope effectively becomes a large telephoto lens. Imaging objects with a camera minus its lens through a telescope without an eyepiece is called prime focus photography. A 200 mm *f*/10 SCT can project the entire half-degree-wide lunar disk onto a single 35 mm film frame. Prime focus is the best method to capture a sequence of lunar phases, a close planetary appulse or a small comet. Higher magnifications will be produced if an eyepiece is used to project the image directly onto the film. Eyepiece projection is capable of capturing good views of the Jovian system with the four bright Galilean satellites, or for revealing detail within the area surrounding a comet's nucleus.

Digital cameras

Digital cameras are versatile, and can be adjusted to suit the conditions. Unlike conventional photography, an image can be instantly reviewed on the camera's screen to assess its quality, and can be removed from the memory if it falls short. Once transferred to a computer, the digital images can be manipulated in numerous ways with an image program to enhance color and detail. The higher the megapixel rating of a digital camera – the number of light-sensitive pixels on its

◀ *Digital cameras can provide pleasing views of the Moon, even if the image is taken by simply holding the camera to the eyepiece of an undriven telescope. This image was taken by the author on December 27, 2004, with a Pentax Optio S30 through a 200 mm SCT.*

◀ *Camcorders are capable of capturing dynamic events like eclipses and bright stellar and planetary occultations by the Moon. Individual camcorder frames will not withstand great enlargement, but they can be stacked to produce detailed images. This camcorder still frame from a movie made by the author using a 127 mm MCT afocally shows Saturn (bottom right) after its occultation by the crescent Moon on April 16, 2002.*

tiny light-sensitive CCD chip – the better its resolution will be. Low-end digicams have small CCD chips with resolutions as low as 640 × 480 pixels (less than a third of a megapixel); individual images taken with such cameras may appear somewhat grainy when enlarged. High-end digital cameras have resolutions greater than 3.1 megapixels, and come packed with a host of superb features, such as the ability to capture movie sequences, take time exposures and compensate for low light-level conditions.

Since most digicams have non-removable lenses, they can only be used afocally through a telescope, so are prone to the same problems of image vignetting as ordinary film cameras. This is easier to overcome with a digicam, however, since most have viewscreens and an optical zoom facility which allows closer, unobstructed views of the subject. Once the limit of optical zooming is reached, the digicam's "digital zoom" comes into play. This ought to be avoided, since images captured in this mode gain nothing in terms of quality, and suffer from imaging artifacts.

Camcorders

Camcorder imaging is dependent on the same afocal principles as conventional photography or digicam imaging. The problems of eyepiece selection and vignetting are present, and their solutions are the same as with digital photography (above). By downloading camcorder footage onto a computer, images can be grabbed individually (at low resolution) or digitally stacked with an astronomical imaging program to produce detailed, high-resolution images. The same method is used to produce high-resolution webcam images. Digital video editing temporarily consumes a great amount of hard drive memory and occupies much of the average computer's resources.

Webcam imaging

High-resolution lunar and planetary images can be obtained with quite modest equipment – a converted webcam attached to a small telescope is capable of producing images that would have been the envy of conventional astrophotographers a couple of decades ago. Webcams cost considerably less than dedicated astronomical CCD cameras, yet their performance is on a par when it comes to imaging the Moon and brighter planets. Astronomical CCDs are excellent for imaging fainter Solar System objects, such as the outer planets, distant comets and asteroids.

Any off-the-shelf webcam can be used to image the Moon and the brighter planets. Webcams are often used without their own lens, positioned at the telescope's prime focus, but they can also be used afocally, with both their own lens and the telescope eyepiece in place. CCD chips are highly sensitive to infrared (IR) light, and the webcam's own lens assembly contains an IR blocking filter. If the webcam is used without its lens, a perfectly clean focus through a refractor will not be possible since IR is focused differently from visible light. This problem can be solved by placing an IR blocking filter into the light path before it enters the webcam.

Any telescope and webcam combination will produce a pleasing image of the Moon, but the whole lunar disk is too large to be captured at prime focus. Even using a wide-field, 80 mm, *f*/5 refractor, the entire Moon can only be squeezed into the field if a ×2 focal reducer is used. Prime focus images of the planets are quite small, taking up only a small area of the CCD chip. To produce the best resolution, the image of a planet needs to cover as large an area of the CCD as possible – this can be achieved by using a Barlow lens with a magnifying power of ×2 or ×3.

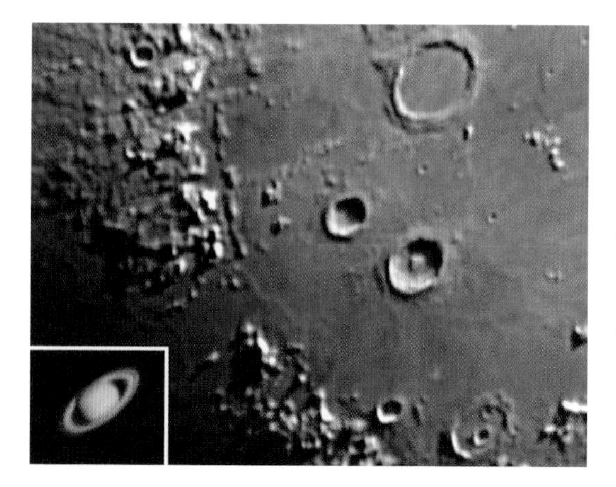

◀ *Comparison between the Moon's Mare Imbrium region and Saturn (inset) imaged with a Philips ToUcam webcam within minutes of each other on October 17, 2003, through a 250 mm Newtonian; they are shown to the same apparent scale.*

Focusing a webcam is achieved by viewing the image on the laptop screen while the focus is being adjusted. An excellent aid to focusing is the Hartman mask, a whole-aperture cover with three equidistant holes in it. An unfocused star will appear as three separate images, and perfect focus is achieved by bringing all three dots to a common point. Electric focusers are highly desirable, since they allow very fine tuning of an image without having to touch the telescope.

Short video sequences can be recorded using the software provided by the webcam manufacturer and stored as video files on the computer's hard drive. Using a USB camera and imaging at 640 × 480 resolution, the optimum imaging rate is ideally limited to a maximum of five frames per second in order to avoid the more obvious artifacts of compression. While the Moon can often be successfully imaged in full auto mode, the bright planets require the settings to be adjusted to produce the best balance of exposure, brightness, contrast and color. By limiting the frame rate to a maximum of five frames per second, a video duration of 10 seconds will produce a total of 50 frames in a video exceeding 1 Mb in size. It is not unusual for some dedicated lunar and planetary imagers to use a single video clip of hundreds or even thousands of frames to produce a single final image.

Image-editing software, such as the freeware Astrostack or RegiStax, is used to extract individual images from the video. A well-aligned and smoothly operating equatorial mount is essential to good webcam imaging: a misaligned mount will produce image drift in even short capture sequences, which produces problems when attempting to align and stack the images. Video files are chopped into individual image frames by the processing software, and the quality of each frame is automatically assessed to gauge its quality. The best images selected are automatically stacked together by the software and combined to remove noise (thus improving the signal-to-noise ratio) and eliminate imaging artifacts. The stacked image can then be processed to bring out detail. Unsharp masking is one of the most useful tools for achieving a contrasty, clean-cut image. It is important to avoid too much image processing, as it can produce spurious artifacts, and fine tonal detail may be lost.

3 · MERCURY

A small, rocky world

Mercury, the closest planet to the Sun, is a small world with a solid rocky surface. Measuring 4880 km in diameter, it is nearer the size of the Moon than the Earth. Indeed, two planetary satellites are actually slightly larger than Mercury – Jupiter's Ganymede and Saturn's Titan – but Mercury is twice as massive as either.

Spaceprobes have revealed that Mercury looks remarkably like the Moon. Its surface is heavily cratered, and it bears a striking resemblance to the lunar southern uplands. Large expanses of maria – relatively flat, smooth plains of dark lava which flooded the floors of large asteroid impact basins – are absent on Mercury. Mercury's terrain consists of ancient, highly eroded craters, interspersed with younger cratered areas, hummocky plains, ridges, scarps and fractures. Like the Moon, most of these craters are thought to have been excavated by asteroidal impact between 4.5 and 3 billion years ago. Some small smooth plains on Mercury may have been produced by much more recent volcanism.

A gigantic asteroid impact basin called Caloris Planitia, measuring 1300 km across, is Mercury's largest single feature. Huge shock waves propagating away from the focus of the impact set up a series of concentric faults in the crust. After the impact, the crust slowly adjusted, steep scarps and ridges appeared along the fault lines, and there was a partial in-filling of the Caloris basin by lava.

Some of Mercury's younger craters are surrounded by sheets of ejected material and bright spindly rays, but these tend to cover a smaller area in relation to their parent craters than their lunar counterparts. Mercury's gravitational pull of more than twice that of the Moon explains this fact – material thrown out by an impact will be deposited closer to the impact site on planets with a higher gravity.

Plate tectonics have never taken place on Mercury, but that's not to say that its solid crust has not been

MERCURY: DATA	
Globe	
Diameter	4880 km
Density	5.43 g/cm^3
Mass (Earth = 1)	0.0553
Volume (Earth = 1)	0.0562
Sidereal period of axial rotation	58d 15h 30m
Escape velocity	4.4 km/s
Albedo	0.1
Inclination of equator to orbit	0°
Surface temperature	100–800 K
Surface gravity (Earth = 1)	0.38
Orbit	
Semimajor axis	0.387 AU = 57.91 × 10^6 km
Eccentricity	0.206
Inclination to ecliptic	7° 00′
Sidereal period of revolution	87.969d
Mean orbital velocity	47.89 km/s
Satellites	0

► *NASA's Mariner 10 spaceprobe photographed Mercury on three occasions during 1974 and 1975. This image shows the 350-km-long Discovery Scarp, a huge cliff formed along a fault line in the planet's thick crust.*

subject to tremendous deforming forces. After the planet's formation, its metallic core cooled and contracted. Since Mercury's core is proportionately huge – perhaps in excess of 70% the planet's diameter – its shrinkage had startling repercussions, with the solid crust deforming and developing wrinkles like the skin of an old apple.

Mercury has only the merest trace of an atmosphere – a tenuous mix of various gases produced by the constant bombardment of its surface by solar particles and meteorites. Barely distinguishable from a vacuum, and millions of times too thin for any weather to be produced, Mercury's atmosphere cannot be detected from the Earth through backyard telescopes. Without an appreciable atmosphere to distribute heat around the planet, there is a great temperature difference between Mercury's daytime and night-time hemispheres. Its Sun-facing hemisphere is heated to around 350°C at the equator, while the night side plunges to around −180°C – the most extreme temperature variation found on any planet.

Orbit

Mercury orbits in an elliptical path that takes it as far as 69,815,900 km from the Sun at aphelion to as close as 46,003,500 km at perihelion. Its orbital plane is tilted by just over 7° to the plane of the ecliptic. Out of all the planets, only distant little Pluto has a more eccentric and more steeply inclined orbit.

Instead of returning to the same point in space after completing each orbit, the point of Mercury's perihelion advances. Viewed from above, over time the orbit describes a "rosette" pattern like a spirograph drawing. The amount of perihelion precession observed cannot be explained by classical Newtonian physics, and astronomers in the 19th century suggested that the gravitational pull of a planet nearer the Sun might be responsible. Not only was the orbit of this supposed intramercurial planet calculated, but it was also given a name – Vulcan – and was even supposedly observed on a number of occasions, both near the Sun and in transit across the solar disk.

Such romantic notions of an elusive inner planet were quashed in the early 20th century, when Mercury's perihelion precession was shown

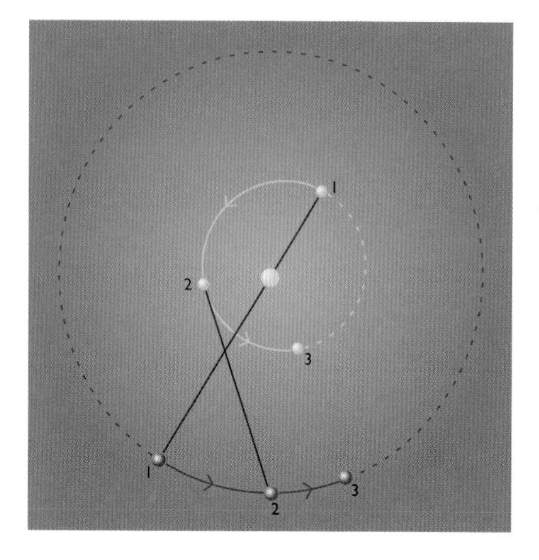

◀ *The orbits of Mercury (pale blue) and the Earth (dark blue) in early 2007, drawn to scale. Mercury is at superior conjunction on January 7, 2007, when Earth and Mercury are at point 1 in their orbits. When the planets are at point 2 in their orbits, Mercury is at its greatest elongation east (18° from the Sun), on February 7, 2007. At points 3, Mercury is at inferior conjunction, on February 23, 2007.*

to be a consequence of general relativity. The planet's proximity to the gravitational field of the Sun, and the large degree to which the space around the Sun is strongly curved, brings into play relativistic effects in Mercury's orbit that can be readily observed and measured.

Rotation

Under the impression that its rate of spin was kept in check by the Sun's gravity, astronomers used to think that Mercury was in a captured rotation, the time it took to complete one revolution on its axis being the same as the time it took to make one orbit around the Sun. Just as our own Moon is in a captured rotation around the Earth, and keeps the same face turned toward us throughout its entire orbit, Mercury was thought to constantly present the same hemisphere toward the Sun. This notion was reinforced by visual observations, which appeared to confirm that the vague surface features observed on Mercury were always in the same position with respect to the Sun.

Radar observations in the mid-1960s showed that Mercury's period of rotation does resonate with its orbit around the Sun, but not on a 1:1 basis like the Moon and Earth. Mercury has a 3:2 spin–orbit resonance, which means that it rotates on its axis three times for every two revolutions around the Sun. Its rotational axis is exactly perpendicular to the plane of its orbit – a situation unique in the Solar System. Mercury takes 59 days to make one complete turn on its axis (rotating from west to east), which is exactly two-thirds of the time it takes to make its 88-day orbit around the Sun. An area near the planet's equator is illuminated by the Sun for periods lasting 88 days.

Apparitions

Observed from the Earth, Mercury never strays very far from the Sun. As a consequence, it is not possible to observe Mercury in a dark sky. Even at its furthest elongation from the Sun, it appears set against a twilit sky. Due to Mercury's rapid motion, it is usually only visible with the unaided eye for a few weeks during each elongation in the most favorable circumstances. At a favorable eastern elongation, Mercury only becomes visible with the unaided eye around half an hour after sunset. Depending on the flatness of the horizon and the clarity of the atmosphere, it can remain visible in a darkening twilit sky for up to an hour until it fades near the horizon. During a favorable morning apparition, Mercury can be picked up with the unaided eye around an hour and a half before sunrise; it fades from view in brightening skies when it is a dozen or so degrees above the horizon, around half an hour before sunrise. Future elongations of Mercury are given in Table 7.

Not all elongations of Mercury are equally as favorable to view from any one point on the Earth. Due to the planet's orbital eccentricity, greatest elongations from the Sun vary between a maximum of 28° (when it is near aphelion) and a minimum of 18° (near perihelion). The best views of Mercury are to be had when it is at its greatest elongation from the Sun and placed as high as possible above the horizon at sunset or sunrise. From northern temperate regions such as the UK, Europe and much of North America, such circumstances arise during favorable eastern elongations in springtime, or when the planet is at a maximum western elongation in an autumn morning sky. The opposite applies to observers in the southern hemisphere.

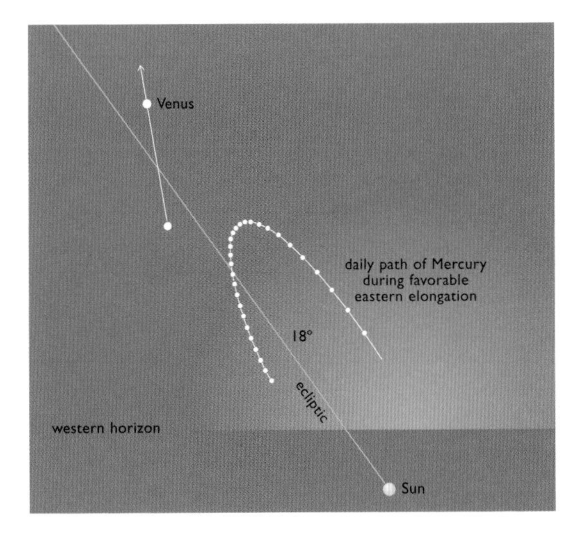

▶ *Typical daily path of Mercury above the western dusk horizon during a favorable eastern elongation. This example uses the elongation of early 2007, plotted between January 21 and February 18 , as viewed from Birmingham, UK. Mercury's maximum eastern elongation of 18° occurs on February 7.*

Transits

Mercury's synodic period – the interval between inferior conjunctions – is around 116 days. Since its orbit is inclined by about 7° to the plane of the ecliptic, Mercury rarely moves directly in front of the face of the Sun, instead passing well to the north or south of the Sun at inferior conjunction. Transits take place when Mercury moves directly across the line joining the Sun and the Earth. No fewer than 14 transits of Mercury took place during the 20th century. The combined orbital circumstances of Mercury and the Earth cause transits to happen only during the months of May and November. May transits recur every 13 and 33 years, while November transits recur at intervals of 7, 13, and 33 years. Since Mercury is considerably closer to the Earth during May transits, it appears as a black disk some 12 arcseconds in apparent diameter, while its disk is just 10 arcseconds across when viewed during November transits. It is, however, far too small to be seen with a protected naked eye, and must instead be viewed either using a telescope with a full aperture filter or by projecting its image onto a shielded white card to view safely.

From its first appearance as a tiny indentation on the Sun's limb (just after first contact), Mercury can take as little as two minutes to enter fully upon the Sun (second contact), depending on the angle at which it approaches the solar limb. In the past, some observers have claimed a "black drop" effect, where the planet briefly appears to remain attached to the Sun's limb after

TABLE 6 FUTURE TRANSITS OF MERCURY

Date	Time at mid-transit (UT)
2006 Nov 08	21:41
2016 May 09	14:57
2019 Nov 11	15:20

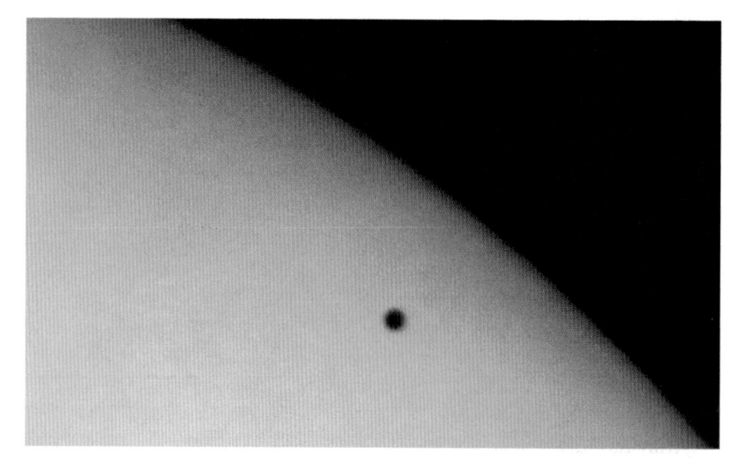

▼ *Mercury in transit across the Sun on May 7, 2003, imaged by Jamie Cooper.*

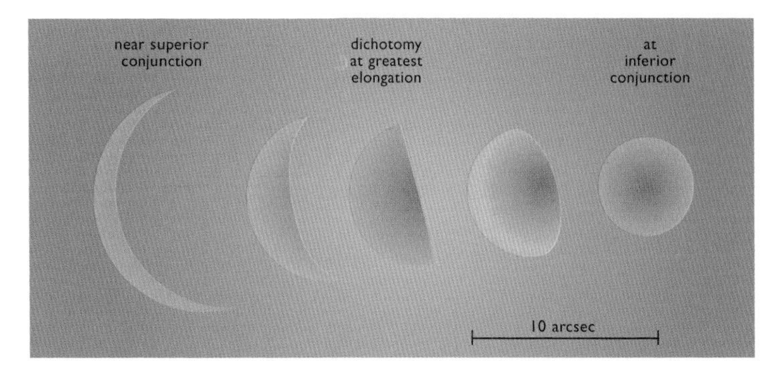

▲ *The phases of Mercury, drawn by the author to the same angular scale.*

second contact. Such an effect is caused by a combination of atmospheric seeing and the instrument's own optics. Since Mercury has no atmosphere to speak of, none of the various phenomena observable during a transit of heavily cloaked Venus are visible. Depending on the size of the chord traced by the planet's movement against the solar disk, transits can last from a few minutes (for a grazing transit at the Sun's limb) to more than four hours. Table 6 gives the dates of future transits of Mercury.

Mercury's phases and its appearance through the telescope

Mercury can be observed telescopically for a maximum of around five weeks during favorable apparitions. Being an inner planet, Mercury displays a sequence of phases when viewed through even a small telescope used at a fairly high magnification. At superior conjunction Mercury is fully illuminated, but lies far too close to the Sun for regular observation. As Mercury pulls away from the Sun in the evening skies after superior conjunction, it draws closer to the Earth and its phase begins to decrease. Mercury reaches its brightest magnitude when it assumes a gibbous phase, around 75% illuminated. At its brightest, Mercury shines at around magnitude −0.01, which is just half a magnitude fainter than Sirius, the brightest star. At greatest elongation east, Mercury has reached a half-illuminated phase, its disk around 7 arcseconds in diameter. The planet becomes a crescent, slowly growing in size by a couple of arcseconds until it is lost in the evening twilight as it draws closer to the Sun and its inferior conjunction. At inferior conjunction, Mercury's Earth-facing hemisphere is completely unilluminated – the planet is only visible at inferior conjunction during transits. Afterward, the planet pulls out into the predawn skies, and over the next few weeks displays a phase sequence from crescent, through half-illuminated to gibbous.

Long before its surface had been imaged up close, much had been deduced about the true nature of Mercury's surface from visual observations alone. Mercury's albedo (a measurement of the amount of sunlight it reflects) is 0.1, which is equivalent to the overall albedo of the Moon. Its light curve, from the bright limb to the terminator (the division between the planet's illuminated and unilluminated hemispheres), also equates with that of the Moon. Regional albedo variations are less pronounced on Mercury than on the Moon; while the Moon has a patchwork of dark maria and brighter highlands, Mercury's overall shading is far more subtle.

A few observers using excellent instruments under the most favorable conditions have claimed to have discerned craters along Mercury's terminator, where the low angle of illumination causes topographic features to cast shadows. Although some of these claims may be true, such features would be exceptionally elusive. Using a 100 mm telescope at high magnifications under good seeing conditions, shading along Mercury's terminator is often visible, along with some ill-defined albedo shadings spreading across the planet's illuminated face.

TABLE 7 FUTURE ELONGATIONS OF MERCURY

Elongation	Date	Distance from Sun	Magnitude
Eastern	2006 Feb 24	18.2°	−0.02
Western	2006 Apr 8	27.3°	+0.06
Eastern	2006 Jun 20	24.2°	+0.07
Western	2006 Aug 7	19.2°	+0.03
Eastern	2006 Oct 17	24.2°	+0.02
Western	2006 Nov 25	19.2°	−0.03
Eastern	2007 Feb 7	18.2°	−0.03
Western	2007 Mar 22	27.3°	+0.05
Eastern	2007 Jun 2	23.2°	+0.07
Western	2007 Jul 20	20.2°	+0.05
Eastern	2007 Sep 29	26.3°	+0.03
Western	2007 Nov 8	19.2°	−0.03
Eastern	2008 Jan 22	18.2°	−0.03
Western	2008 Mar 3	27.3°	+0.03
Eastern	2008 May 14	21.2°	+0.06
Western	2008 Jul 1	21.2°	+0.06
Eastern	2008 Sep 11	26.3°	+0.04
Western	2008 Oct 22	18.2°	−0.03
Eastern	2009 Jan 4	19.2°	−0.03
Western	2009 Feb 13	26.3°	+0.02

Elongation	Date	Distance from Sun	Magnitude
Eastern	2009 Apr 26	20.2°	+0.04
Western	2009 Jun 13	23.2°	+0.07
Eastern	2009 Aug 24	27.3°	+0.05
Western	2009 Oct 6	17.2°	−0.03
Eastern	2009 Dec 18	20.2°	−0.03
Western	2010 Jan 27	24.2°	+0.01
Eastern	2010 Apr 8	19.2°	+0.03
Western	2010 May 26	25.3°	+0.07
Eastern	2010 Aug 7	27.3°	+0.06
Western	2010 Sep 19	17.2°	−0.01
Eastern	2010 Dec 1	21.2°	−0.02
Western	2011 Jan 9	23.2°	−0.00
Eastern	2011 Mar 23	18.2°	+0.01
Western	2011 May 7	26.3°	+0.07
Eastern	2011 Jul 20	26.3°	+0.07
Western	2011 Sep 3	18.2°	+0.00
Eastern	2011 Nov 14	22.2°	−0.00
Western	2011 Dec 23	21.2°	−0.02
Eastern	2012 Mar 5	18.2°	−0.01
Western	2012 Apr 18	27.3°	+0.06
Eastern	2012 Jul 1	25.3°	+0.07
Western	2012 Aug 16	18.2°	+0.02
Eastern	2012 Oct 26	24.2°	+0.01
Western	2012 Dec 4	20.2°	−0.03
Eastern	2013 Feb 16	18.2°	−0.02
Western	2013 Mar 31	27.3°	+0.05
Eastern	2013 Jun 12	24.2°	+0.07
Western	2013 Jul 30	19.2°	+0.04
Eastern	2013 Oct 9	25.3°	+0.02
Western	2013 Nov 18	19.2°	−0.03
Eastern	2014 Jan 31	18.2°	−0.03
Western	2014 Mar 14	27.3°	+0.04
Eastern	2014 May 25	22.2°	+0.07
Western	2014 Jul 12	20.2°	+0.06
Eastern	2014 Sep 21	26.3°	+0.03
Western	2014 Nov 1	18.2°	−0.03
Eastern	2015 Jan 14	18.2°	−0.04
Western	2015 Feb 24	26.3°	+0.03
Eastern	2015 May 7	21.2°	+0.05
Western	2015 Jun 24	22.2°	+0.07
Eastern	2015 Sep 4	27.3°	+0.05
Western	2015 Oct 16	18.2°	−0.03
Eastern	2015 Dec 29	19.2°	−0.03

Recording Mercury

A circular template 50 mm in diameter is commonly used for drawings of Mercury. The planet's phase should be drawn in first, and any detail then added. Sharply defined features are unlikely to be visible, and any shading is likely to be faint and indistinct. Terminator shading, if any, is then drawn in, along with any apparent irregularities visible along the terminator. Sometimes, when the planet is at a crescent phase, its southern cusp appears somewhat blunted, and its northern cusp is somewhat brighter and more sharply delineated. Occasionally, brighter patches can be seen on the disk – these are likely to be bright ray systems surrounding Mercury's more recent impact craters. When observing Mercury visually in a dawn or dusk sky, the contrast can be increased by using a red filter.

Intensity estimates

A line drawing of the planet with intensity estimates provides a good reference. The scale of intensities is a relative one:

0	Brilliant white
1	Bright areas
2	General tone of the disk
3	Elusive, low-contrast shadings at the limit of visibility
4	Distinct shadings
5	Prominent dark shadings

CCD images are capable of revealing detail on Mercury, but the planet's small apparent diameter requires a hefty magnification to capture at prime focus. Again, filters will be of considerable use, particularly red filters, in combination with an IR blocking filter.

◀ *Mercury observed by the author through a 300 mm Newtonian on May 4, 2002, around the period of its greatest elongation east.*

4 · VENUS

An Earth-sized world

A terrestrial planet measuring 12,104 km in diameter, Venus is almost as large as the Earth. Its substantial atmosphere hosts such a thick blanket of clouds that its surface can never be observed visually.

Radar studies conducted from the Earth and by spaceprobes orbiting Venus have mapped its veiled surface in considerable detail. Much of the planet is covered by great rolling plains. Rising from these undulating regions are a number of elevated plateaux, from which rise two major mountain ranges. Ishtar Terra, in Venus' northern hemisphere, is around the size of Australia. Towering above its eastern edge is the planet's highest mountain, Maxwell Montes; at 12 km high, it is more than three kilometers higher than Mount Everest. Aphrodite Terra, around the same size as Africa, is the largest of Venus' mountain ranges. Located south of Venus' equator, this vast highland plateau stretches half-way around the planet. An enormous valley, Diana Chasma, winds through southern Aphrodite; in places, Diana Chasma is 280 km wide and 4 km deep.

Although Venus' crust has displayed localized tectonic activity, with extensive faulting and volcanism, it is effectively a single plate, rather than a jigsaw assembly of plates like the crust of the Earth. Venus' sprawling mountain plateaux are the result of hundreds of millions of years of intensive volcanism occurring at specific points in a static crust; this has enabled vast amounts of material to extrude on to the surface and sprawl across the local terrain.

Unlike most other solid worlds in the Solar System, there's a notable dearth of impact craters on Venus. There is no doubt that through the eons Venus has had its fair share of asteroidal bombardment. Just like the Earth, however, Venus' dense atmosphere has ameliorated the effects of this bombardment. On encountering a dense atmosphere, impactors burn up, fragment or disintegrate

VENUS: DATA	
Globe	
Diameter	12,104 km
Density	5.20 g/cm^3
Mass (Earth = 1)	0.8149
Volume (Earth = 1)	0.8568
Sidereal period of axial rotation (retrograde)	243d 0h 30m
Escape velocity	10.4 km/s
Albedo	0.65
Inclination of equator to orbit	177° 20′
Surface temperature	750 K
Surface gravity (Earth = 1)	0.90
Orbit	
Semimajor axis	0.723 AU = 108.2×10^6 km
Eccentricity	0.007
Inclination to ecliptic	3° 24′
Sidereal period of revolution	224.701d
Mean orbital velocity	35.02 km/s
Satellites	0

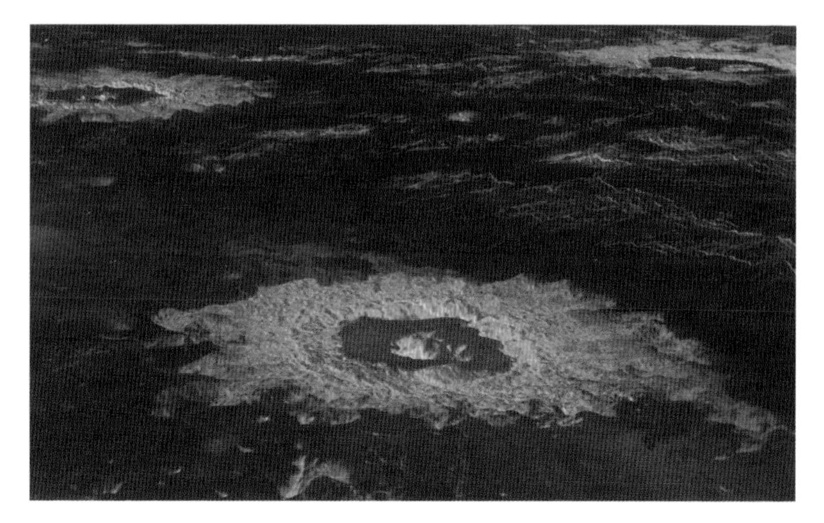

▲ *Using data obtained by radar onboard the Magellan spaceprobe, the appearance of Venus' amazing topography can be simulated, as in this oblique view of three sizable impact craters.*

altogether, sometimes exploding in the atmosphere before they reach the ground. Those bodies large enough to make it down to the surface produce an impact crater – generally speaking, the larger the impactor and the higher its velocity, the bigger the crater produced. Since Venus' atmosphere is dynamic and its surface volcanically and tectonically active (on a local scale), impact features are quickly eroded, covered by lava or deformed, or modified by a combination of all three processes. This explains why there are so few impact craters visible on Venus' surface today – those that are visible are very young features in terms of the age of Venus, probably less than 500 million years old.

Hell in disguise

From the perspective of human exploration, Venus is one of the least inviting places in the entire Solar System. The bright cloud layer preventing its surface from being viewed is stacked between 50 and 70 km above the surface. The clouds themselves are composed of sulfuric acid, and the atmosphere is largely composed of carbon dioxide gas. At Venus' surface, air pressure is 90 times that on Earth at sea level, and the temperature on all parts of the planet, day and night, is maintained at around 460°C – about 200°C hotter than a kitchen oven at its maximum setting. Venus' high temperatures are caused by a runaway greenhouse effect. While much of the solar energy reaching Venus is reflected by its bright cloud tops, some penetrates through the atmosphere and down to the surface, where it is absorbed and then re-radiated as heat. Most of the heat emitted by the surface cannot escape from the atmosphere, and it is reflected back down to the surface by the clouds. Venus' searingly high surface temperatures are maintained in this relentless manner, with nothing to break the cycle.

Orbit and rotation

Revolving about the Sun at an average distance of 108,200,000 km from the Sun in a period of 224.7 days, Venus has the most circular orbit of any of the planets. It approaches us closer than any other planet, sometimes coming as near as 38,151,000 km when the Earth is near perihelion at the time of Venus' inferior conjunction. Venus' orbit is inclined around 3.4° to the ecliptic, which means that it is rarely exactly in line with the Sun and the Earth at inferior conjunction. Instead, Venus usually passes a little to the north or south of the Sun.

Since Venus' surface is permanently hidden from view, astronomers had no way of knowing how fast the planet rotated on its axis until the advent of radar studies in the mid-20th century. Venus was found to have a slow east-to-west spin – a retrograde spin, opposite to the axial rotations of most other planets. It takes 243 days to turn once on its axis – somewhat longer than its 224.7-day period of revolution around the Sun – which means that daytime on a typical part of equatorial Venus lasts around eight weeks. Despite the fact that Venus is, on average, more than 40 million km closer to the Sun than the Earth, daytime there is a gloomy affair. So little sunlight percolates through the thick clouds that light levels at midday under a Venusian summer Sun are comparable to a heavily overcast midwinter day in London or Seattle.

Apparitions

With the exception of a few rare transient astronomical phenomena, such as a really brilliant comet or a swiftly moving fireball, Venus is the third-brightest celestial object. Second only in brilliance to the Sun and the Moon, its dazzling presence high in the evening or morning skies often attracts the attention of non-astronomers, most of whom imagine it to be the lights of a relatively nearby aerial object.

Within a month or so following superior conjunction, as Venus edges east of the Sun, it becomes visible with the unaided eye in the evening skies. Venus shines with a steady, brilliant white hue, and even when fairly low near the horizon may not display any signs of scintillation caused by atmospheric turbulence. Following superior conjunction, it takes Venus around seven months to reach maximum eastern elongation, between 45° and 47° from the Sun. This is a respectable angular separation, nearly twice the best elongations of Mercury from the Sun, and easily enough distance for Venus to be seen against a truly dark evening sky, several hours after sunset.

Some apparitions of Venus are more favorable than others, depending on how high the planet is above the horizon at sunset. Maximum eastern elongations are most favorable from northern temperate latitudes during the spring, when the planet can be as high as 40° above the western horizon at sunset and observable for more than four hours in darkening evening skies. For observers in northern temperate

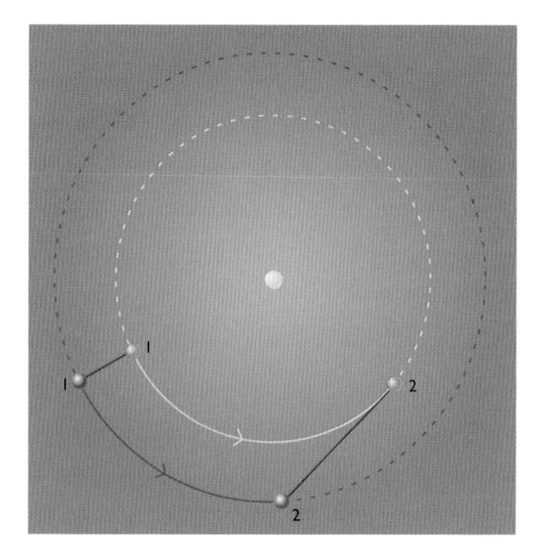

◀ *The paths of Venus (yellow) and the Earth (blue) drawn to scale, from inferior conjunction to greatest elongation west. When Earth and Venus are at points 1 in their orbits, Venus is at inferior conjunction, on August 16, 2015. When the planets are at points 2, Venus is at its greatest elongation west (47°), on October 26, 2015.*

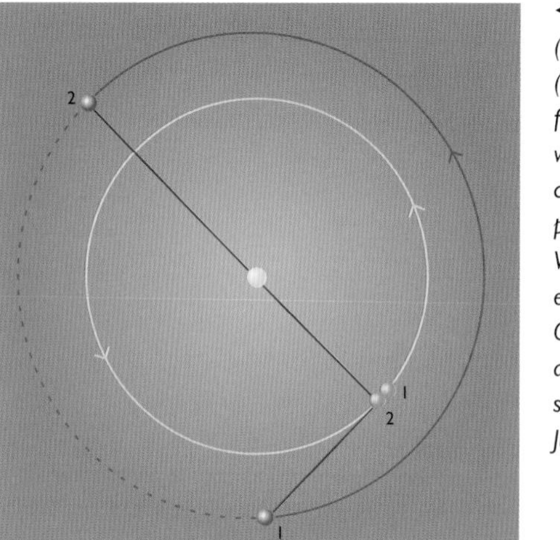

◀ *The paths of Venus (yellow) and the Earth (blue) drawn to scale, from greatest elongation west to superior conjunction. When the planets are at points 1, Venus is at its greatest elongation west (47°), on October 26, 2015. When at points 2, Venus is at superior conjunction, on June 6, 2016.*

regions, unfavorable eastern elongations take place during the autumn, when the angle made by Venus, the setting Sun and the horizon is at its smallest. On these occasions, Venus can be less than 10° above the southwestern horizon at sunset. With such a low altitude, the telescopic image of Venus is liable to suffer greatly from the effects of atmospheric turbulence. Similarly, observers in southern locations such as Australia and New Zealand experience their most favorable eastern elongations during the southern spring, when Venus is visible high above the western horizon after sunset. Maximum eastern elongations during the southern autumn are poor, with the planet at a low altitude.

Venus' journey back toward the Sun is accomplished much faster than its outward leg, taking just 10 weeks or so to move from maximum eastern elongation to inferior conjunction. Venus' maximum brightness in the evening skies, when it gleams a brilliant magnitude −4.3, is reached around 36 days before inferior conjunction, at an elongation of around 39° east of the Sun, when it is a crescent phase a little more than 25% illuminated. On these occasions, it is possible to see Venus in a clear, haze-free sky during full daylight using the unaided eye. To have the best chance of viewing Venus with the naked eye, it is best to hide the Sun from direct view using a local object, such as the edge of a roof or the side of a building. While looking at a bland area of sky, the eyes have no reference object in the far distance upon which to focus, so locating a faint pinprick of light in a bright sky can be a challenge. One technique is to scan the far horizon, and then quickly turn one's gaze to the area of sky in which Venus is thought to be positioned; by doing this, the eyes are likely to retain their distance focus for a short while.

Following inferior conjunction, Venus proceeds to make its presence known to the west of the Sun in the predawn skies. As with evening apparitions, maximum brightness takes place around 36 days into the apparition when Venus is around 39° east of the Sun, a crescent phase more than a quarter illuminated and shining at magnitude −4.3. It takes Venus around 10 weeks to reach maximum elongation west of the Sun. For observers in northern temperate latitudes, maximum western elongations are at their most favorable when they occur during the autumn, when Venus rises above the eastern horizon almost five hours before the Sun. Maximum western elongations that take place during the spring are poor, with Venus barely visible above the southeastern horizon before dawn. Similarly, Venus is most favorably placed for southern-hemisphere observers when its maximum western elongation occurs during the southern autumn (a morning object). Venus' journey back toward the Sun and superior conjunction takes around seven months. Future elongations of Venus are given in Table 9.

Transits

When Venus moves directly between the Sun and the Earth at inferior conjunction, it can be seen as a black circle in transit across the solar disk. Despite covering just one-thousandth the area of the Sun during transit, Venus' apparent angular diameter is so large (a little under 1 arcminute across) that it can be seen without optical instruments by those with keen eyesight, provided that the eyes are safely shielded from direct sunlight by using a proper solar filter. Special solar eclipse eye shades use a thick layer of aluminized mylar to reflect the Sun's rays, preventing most of its light, heat and ultraviolet radiation from reaching the eyes; these shades can be used for brief but safe views of the Sun. Under no circumstances should such eye shades be worn while observing the Sun through the telescope eyepiece, however, since the intense magnified energy of the Sun will quickly burn through them and cause permanent eye damage, if not blindness. There are only two safe ways to observe the Sun through a telescope – eyepiece projection and whole aperture filtration (see Chapter 15).

Transits of Venus take place every 8, 121.5, 8 and 105.5 years. Only seven of these events have taken place since the telescope was invented in the early 17th century, the first of these having been observed from England in 1639. The last one took place on June 8, 2004, and it proved to be one of the most widely observed astronomical events in history, viewable in its entirety from the UK, Europe, India and most

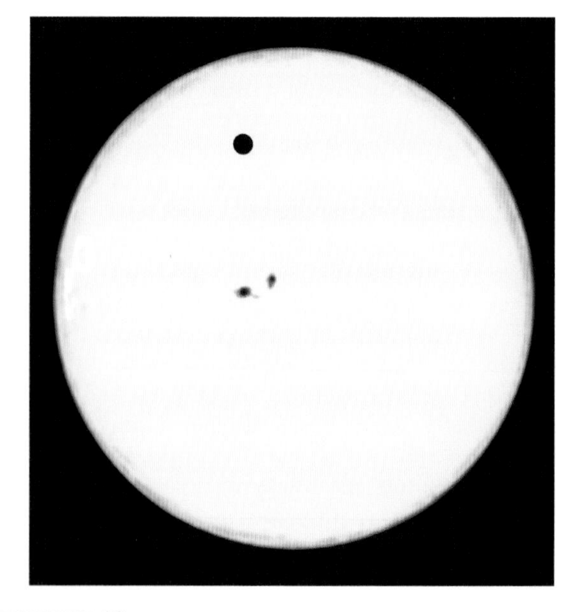

◀ Venus transits the Sun on June 8, 2004, observed by the author at 08:10 UT using a 200 mm SCT with an Inconel full-aperture solar filter.

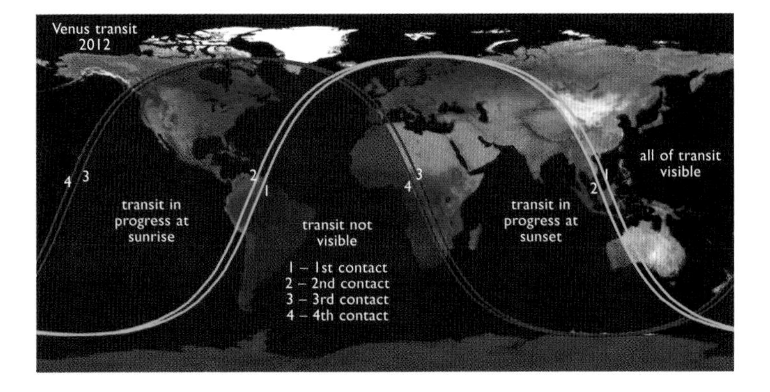

▲ *The June 6, 2012, transit of Venus, showing lines of visibility for all four contacts.*

of Asia. From eastern North America and much of South America, Venus was already on the Sun's disk as the Sun rose, while the transit was still in progress at sunset for observers in Japan and Australia. There is one more opportunity to view a transit of Venus during the 21st century, and it takes place on June 6, 2012. The next pair of transits will occur in 2117 and 2125.

The Venus transit of June 6, 2012

The circumstances for the transit of 2012 are almost a reversal of those of 2004. Observers in northwestern North America, the western Pacific, northeastern Asia, Japan, eastern Australia and New Zealand will have a view of the whole event, from first to last contact. From Hawaii, the Sun will be almost directly overhead when the transit begins and will not set before it ends. From most of North America and northwestern South America, the Sun sets while the transit is still in progress. For observers in central Asia, most of western Europe (including the UK), eastern Africa and eastern Australia, the last part of the transit will be visible as the Sun rises. However, Portugal and much of Spain, western Africa and most of South America will be denied a view of the transit.

First contact occurs when the leading edge of Venus touches the edge of the Sun, at around 22:09 UT (Universal Time). This is not as easy an observation to make as might first be suspected. The observer will need to do some basic research into where on the Sun's edge Venus is due to make its long-awaited entrance. The position angle of Venus' first contact, measured in degrees starting from the northern point of the Sun's disk, eastward around the limb, is 40.7°. Of course, this important information needs to be translated into the orientation

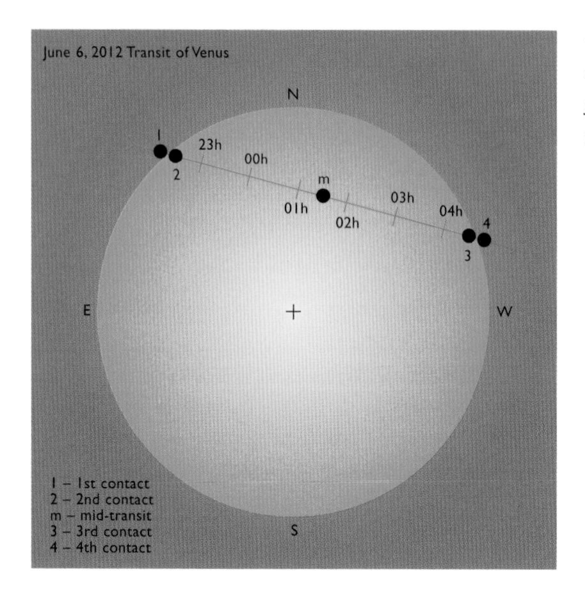

◀ *The transit of Venus across the solar disk on June 6, 2012. Times are given in UT.*

of the Sun as viewed through the telescope eyepiece or on the Sun's projected image. Observed through an astronomical telescope with an adequate full-aperture solar filter, north is at the bottom of the image and east is on the right. Using a diagonal eyepiece, the image is flipped vertically, so that north is at top, but east remains on the right. On an image projected from the eyepiece of an astronomical telescope, north is at the bottom, but east is on the left-hand side. Since the orientation of the Sun slowly swings from east to west during the day, its northern point rotating with respect to true north, the orientation of the Sun will vary depending on the time of observation and the observer's geographical location. It is possible for an inexperienced observer to be watching for Venus' ingress not only on the wrong part of the Sun's limb, but also on the wrong side altogether of the Sun, only to notice the presence of Venus long after first contact, when it has already taken quite a considerable bite out of the northeastern edge of the Sun. Observers using small telescopes and low magnifications will generally notice Venus following its first contact some time – perhaps as much as a minute or so – after observers using high magnifications who are concentrating their attention upon a specific point on the solar limb.

Observers viewing the transit through special solar telescopes with filters tuned to hydrogen-alpha frequencies of light may be able to discern the round black disk of Venus a considerable time prior to first contact as seen in white light. This is possible because in hydrogen-alpha light, the Sun's hot chromosphere is visible. The chromosphere lies above the photosphere (the body of the Sun visible in normal white light), and glowing prominences loop above the Sun's edge amid a diffuse chromospheric glow. Approaching the Sun, Venus blocks out the hydrogen-alpha light from these higher atmospheric features and so may be seen even some time before it makes contact with the Sun's actual chromospheric limb. First contact with the chromosphere precedes first contact with the photosphere by several minutes.

As Venus edges onto the Sun's disk, an unusual phenomenon can be observed. Refraction within Venus' thick atmosphere causes sunlight to be bent around the planet's following limb, producing a narrow arc of light around the edge of Venus. This tiny bow of light beyond the edge of the Sun is fascinating to behold – indeed, for many observers it represents the most memorable visual highlight of the transit.

Moving at the rapid angular velocity of around 4 arcminutes per hour, Venus has fully entered onto the Sun within 18 minutes of first contact, the point of second contact being the exact moment when the following limb of Venus just touches the edge of the Sun. A phenomenon known as the "black drop" has often been reported in previous transits, causing a degree of uncertainty in individual timings of the moment of second contact. The black drop gives the visual impression of a lingering ligament of darkness extending from Venus' following limb to the Sun's edge, making Venus appear like a drop of black ink suspended from the edge of the Sun. The phenomenon is not a real

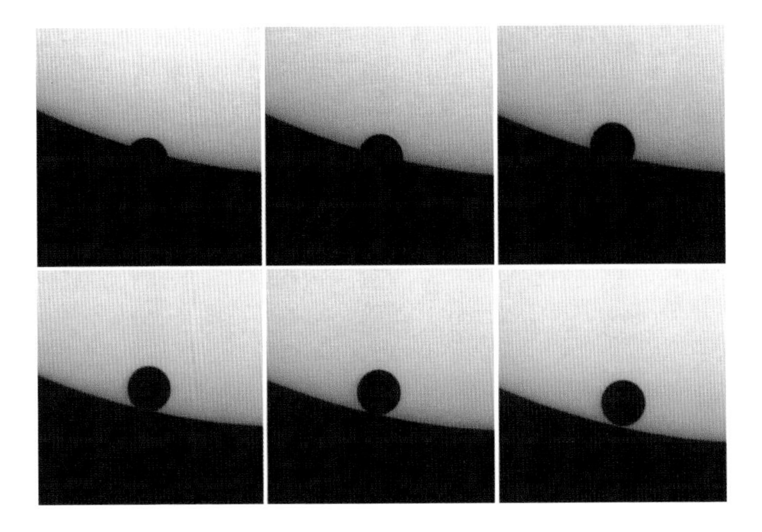

▲ *The ingress stages of the transit of Venus on June 8, 2004, imaged by Jamie Cooper.*

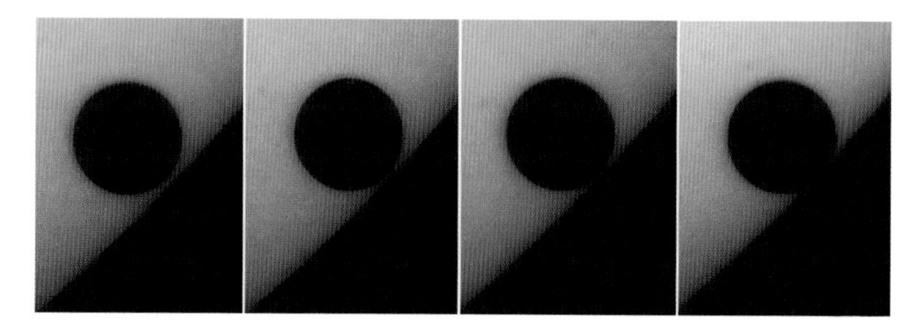

▲ *The egress stages of the transit of Venus on June 8, 2004, imaged by Jamie Cooper.*

event, as it is sometimes described in old literature. Instead, it is caused by poor seeing conditions, poor telescope optics, an inaccurate focus or poor observing skills, or a combination of these factors in various measure. Excellent high-resolution images of Venus at second contact during the 2004 transit show a perfectly clean contact, with no hint of a black drop effect.

Venus reaches the other side of the Sun, after traversing a solar chord some 25 arcminutes long, a little more than six hours later. Note that because the orientation of the Sun changes gradually as it traverses the sky, the observed track of Venus will appear to be a curved line, rather than a straight one. Third contact occurs when the planet's preceding limb touches the Sun's western edge. As the planet leaves the Sun, an arc of refracted sunlight can again be traced around Venus' preceding edge, just as happened after first contact. Fourth contact happens at the very moment when the planet exits the Sun's disk, when the tiny notch disappears and the transit ends for viewers in white light.

The details in Table 8 have been computed for a theoretical view of the transit from the center of the Earth (astronomers call this geocentric data). Actual times will vary slightly depending on the observer's geographic location, but this amounts to no more than a couple of minutes' difference (either earlier or later) at most. Times are given in Universal Time (UT). Mid-transit is at 01:29:28 UT, when Venus is 9 arcminutes from the Sun's northern edge (around one-third of the Sun's apparent diameter).

TABLE 8 DETAILS OF THE TRANSIT OF JUNE 6, 2012

Contact	Time (UT)	Position angle
1st contact	22:09:29	40.7°
2nd contact	22:27:26	38.2°
3rd contact	04:31:31	292.7°
4th contact	04:49:27	290.1°

Locating Venus in daylight

At superior conjunction Venus is a fully illuminated disk just 9 arcseconds in apparent diameter, but the planet is far too close to the Sun to observe through amateur equipment (it is never more than 2 degrees away from the Sun at superior conjunction). As Venus moves east of the Sun it is bright enough to be located through binoculars and telescopes in the daytime skies within three weeks of superior conjunction. To prevent any possibility of eye damage, extreme care must be taken not to accidentally sweep across the raw unfiltered image of the Sun – say, by making sure that the Sun is well hidden behind a solid obstacle before attempting the search. It is possible to follow Venus with relatively modest optical equipment in the daytime skies all the way through the remainder of its elongation east of the Sun and back toward inferior conjunction. Under special circumstances, when Venus is widely separated from the Sun at inferior conjunction (it can be as far as 8 degrees away from the Sun), it is possible to locate the planet through binoculars or a telescope. Again, extreme care must be taken not to accidentally view the Sun directly through any optical instrument, otherwise permanent eye damage is the likely result. Venus can be followed during its western elongations following inferior conjunction in a similar manner.

Attempts to find Venus in the day using binoculars or through a small telescope at a low magnification are best made when the skies are clear and blue, and distant objects can be seen clearly on the horizon. Pollution and atmospheric haze will reduce Venus' brightness and

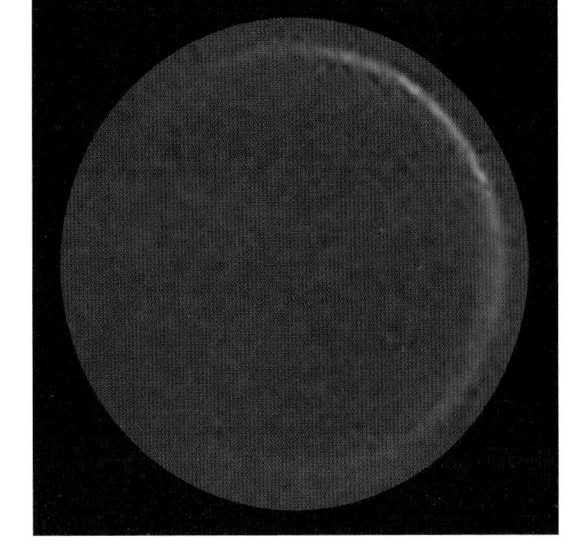

▶ *The extremely slender crescent Venus was imaged during the daylight of June 6, 2004, by Pete Lawrence, just two days before the planet's transit across the Sun. Venus' disk was 57 arcminutes in diameter and just 0.1% illuminated.*

make it harder to find. One of the best celestial stepping stones for finding Venus is the Moon, an object that can usually be found in the daytime sky without any difficulty when it is above the horizon. Every lunar month, the Moon passes within a hand's span of Venus, often coming within a few lunar diameters' distance – these close approaches are known as appulses. A computer program or astronomical almanac enables the observer to find out exactly where Venus is in relation to the Moon at a given time during a daytime appulse, making it easier to locate by just sweeping near to the Moon. Binoculars need to be held in a stable position, by using a rest, camera tripod or binocular mount.

Telescopic observation

When Venus is observed through a telescope in a dark twilit sky, its sheer brilliance can dazzle the eye so much that it can initially be difficult to perceive the planet's phase, let alone any vague atmospheric detail that might be present. Venus' glare is capable of producing some pretty spectacular, though wholly unwanted, special effects. Budget refractors may show Venus surrounded by colored fringes caused by chromatic aberration, and certain types of eyepiece may produce internal reflections and ghosting. Irradiation is another unwanted effect, visible through even the best optics, causing blurring between areas of greatly differing brightness. Irradiation can be produced physiologically by the observer's eye, or as a result of atmospheric turbulence. One way of reducing Venus' glare is to observe when it is high in a bright twilit sky, or even during the daytime.

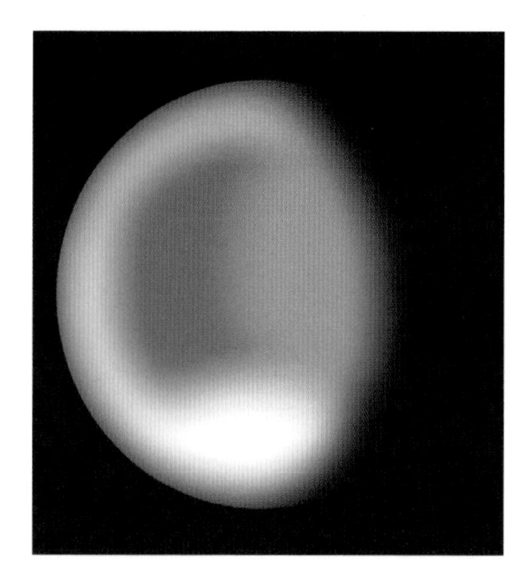

▲ *Cloud features in the atmosphere of a gibbous Venus, observed by the author on February 3, 1996, at 18:05 UT using a 100 mm MCT.*

Using an instrument with accurate setting circles or a computerized drive, Venus can be picked up in broad daylight without much difficulty, providing that it's at a reasonable angular distance from the Sun. Despite this, most observers tend to observe the planet when it is relatively low in the evening or morning sky, when it can be seen with the unaided eye, regardless of the capabilities of the instrument they are using.

▲ *A selection of Venus images taken in ultraviolet light by Jamie Cooper in early 2004 (from left: February 28, April 19, 22 and 24). Cloud features are obvious.*

When observing Venus in twilight, a variable polarizing filter inserted into the eyepiece can be adjusted to produce a satisfyingly glare-free image.

Features in Venus' upper atmosphere are easier to discern than most people imagine. Like any other branch of observational astronomy, the observer needs to attend closely to the object in view and get used to its appearance, and perhaps make a careful observational drawing on each occasion, rather than expect to see features instantly at the most cursory glance through the telescope eyepiece. Venusian cloud features are best observed through a small telescope when the planet is a gibbous phase early in an eastern elongation or late in a western elongation.

Studies of Venus' cloud patterns show that there are constant high-speed winds in the upper atmosphere. Wind speed decreases from Venus' equator toward the poles, causing the appearance of a uniform rotation of cloud features which are blown along by the winds. In the upper atmosphere above the equator, wind speeds of around 360 km/h correspond to a rotation period of four or five days. Wind speeds at the surface of Venus are rather low, amounting to a gentle, though hardly refreshing, breeze.

It is possible to observe the rotation of Venus' upper atmosphere on a daily basis through a telescope as small as 100 mm. During eastern elongations of Venus, features appear to move from the bright limb to the terminator; during western elongations, cloud features appear at the terminator and are carried across the disk to the bright limb. If careful observations are made, the atmospheric patterns will be shown to display this drifting. The most prominent dusky features are usually seen near the terminator, from where they extend and fade, sometimes curving toward the poles. A distinct Y-shaped pattern of clouds is sometimes seen spanning the planet's equatorial region from the terminator toward the limb. Dusky collars bordering brighter polar regions are often seen, giving the appearance of a planet with ice caps like those of Mars. To improve the visibility of Venusian cloud features, blue (Wratten 38A) and violet (Wratten 47) filters are helpful, and so too is yellow (Wratten 12 or 15). Often the planet has a faintly mottled appearance which is very difficult to depict accurately on an observational drawing.

Venusian anomalies

For centuries, observers have noted that the date of Venus' predicted dichotomy (a planet's half-phase) does not always coincide with actual observations. When east of the Sun, Venus often reaches its observed dichotomy some days earlier than the predicted date; at western elongations, observed dichotomy sometimes occurs later than predicted. This phase anomaly is known as the "Schroeter effect" after the lunar and planetary observer Johann Schroeter who noted it in the early 19th century. It is caused by the scattering of sunlight along Venus' terminator; the effects of scattering are more pronounced nearer the planet's edge, where our view is directed through a thicker layer of Venus' atmosphere.

When Venus is a narrow crescent phase, near inferior conjunction, its cusps appear extended around a portion of the limb greater than the area illuminated by direct sunlight. This is another phenomenon caused by the scattering of sunlight within Venus' atmosphere. (Note that it is refraction of sunlight, not scattering, which causes the arc of light observed during transits of Venus.) Around inferior conjunction,

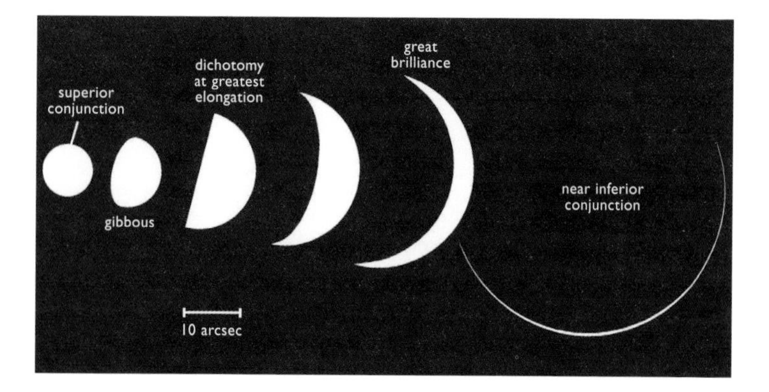

▲ *Phases of Venus drawn to the same angular scale by the author.*

Venus' cusps actually extend so far as to meet on the other side of the planet, producing a curious annular phase. A good separation between Venus and the Sun is required to view such a phenomenon, given the ability to locate Venus safely in daylight through a telescope.

Through the years, a number of reputable planetary observers have reported that Venus' dark side shows a faint illumination when the planet is at a crescent phase. Known as the "ashen light," it has proved to be the most difficult Venusian phenomenon to explain satisfactorily. There seems little doubt that it is a real phenomenon, rather than an optical illusion. Recent theories have suggested a number of possible causes, from the actual glow of Venus' hot surface, to lightning flashes within the Venusian atmosphere.

Recording Venus

Drawings of Venus are made on a 50 mm diameter circular blank. An exaggerated impression of the tonal variation within the planet's cloud features is usually unavoidable, since the features are often so subtle. Intensity estimates of the cloud features can be recorded on a simple line drawing to accompany the shaded observational sketch.

0 Brilliant white
1 Bright areas
2 General tone of the disk
3 Elusive, low-contrast shadings at the limit of visibility
4 Distinct shadings
5 Prominent dark shadings

TABLE 9 FUTURE ELONGATIONS OF VENUS

Elongation	Date	Distance from Sun
Western	2006 Mar 25	46.5°
Eastern	2007 Jun 9	45.5°
Western	2007 Oct 28	46.5°
Eastern	2009 Jan 14	47.5°
Western	2009 Jun 5	45.5°
Eastern	2010 Aug 20	46.5°
Western	2011 Jan 8	47.5°
Eastern	2012 Mar 27	46.5°
Western	2012 Aug 15	45.5°
Eastern	2013 Nov 1	47.5°
Western	2014 Mar 22	46.5°
Eastern	2015 Jun 6	45.5°
Western	2015 Oct 26	46.5°

5 · MARS

Of all the diverse worlds in the Solar System, Mars has been at the center of more scientific attention and people's imaginations than any other. It's not difficult to understand why, because Mars is so similar to the Earth in many ways. It is the only terrestrial planet whose surface features are clearly visible through the telescope eyepiece; it has polar ice caps, and its atmosphere displays weather phenomena, including bright clouds and dust storms. Although we know that advanced biological life forms do not exist on Mars, the question of whether some form of primitive life exists, or has ever existed on Mars, remains as open as ever.

Red planet

Mars is not an enormous planet. Measuring 6795 km in diameter, its dimensions lie neatly between that of the Earth and the Moon. Interestingly, its surface area is around 180 million square km – a little more than the area of dry land on Earth. Viewed with the unaided eye, Mars has a distinctly red hue – the color is due to iron oxide (rust) in the layers of surface dust covering the planet. Mars appears even ruddier through binoculars or a telescope.

Mars' two hemispheres are markedly different in appearance. Extensive dusky tracts cover the southern hemisphere of Mars, while the north appears much brighter, with a few isolated dusky patches. The southern hemisphere is covered with thousands of sizable impact craters, while the north is relatively crater-free. Asteroidal impact has produced most of the craters on Mars, and their appearance has been modified by various weathering processes. Their steep inner slopes often show deep gullies which appear to have been cut by the erosive action of running water, perhaps relatively recently, and sediment from drainage and wind deposition has partly infilled the crater floors. Some

MARS: DATA	
Globe	
Diameter (equatorial)	6795 km
Diameter (polar)	6752 km
Density	3.91 g/cm^3
Mass (Earth = 1)	0.107
Volume (Earth = 1)	0.152
Sidereal period of axial rotation	24h 37m 23s
Escape velocity	5.0 km/s
Albedo	0.15
Inclination of equator to orbit	25° 11′
Surface temperature (average)	220 K
Surface gravity (Earth = 1)	0.38
Orbit	
Semimajor axis	1.524 AU = 227.9 × 10^6 km
Eccentricity	0.0934
Inclination to ecliptic	1° 51′
Sidereal period of revolution	686.980d
Mean orbital velocity	24.13 km/s
Number of satellites	2

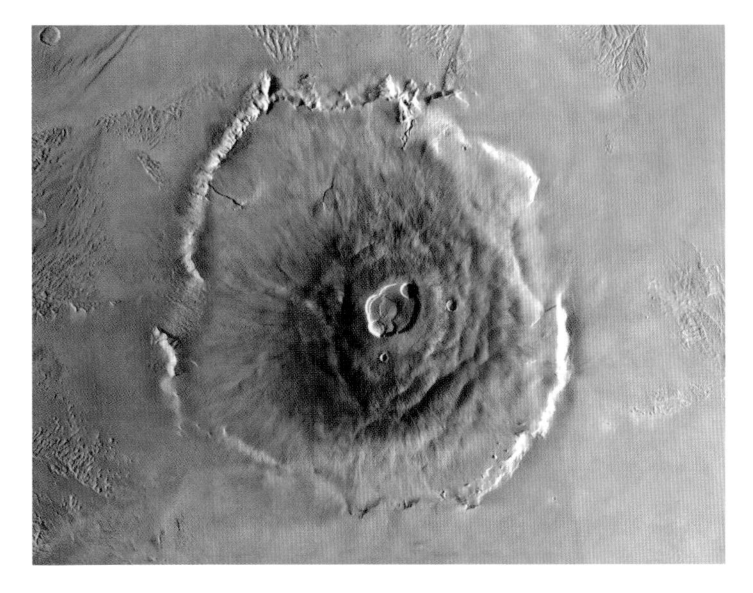

▲ *Rising to a height of more than three times the Earth's loftiest peak, Mount Everest, Olympus Mons is the largest volcano in the Solar System.*

impact craters on Mars are surrounded by peculiar "mudsplash" patterns, indicative of an impact on wet or frozen ground. Far fewer impact craters can be found in the planet's northern hemisphere; many ancient craters have been completely obliterated by lava, which was extruded from volcanic vents in relatively recent geological times. On average, the southern hemisphere is three kilometers higher than the northern hemisphere.

Four immense shield volcanoes dominate the Tharsis region, namely Arsia Mons, Pavonis Mons, Ascraeus Mons and Olympus Mons. With a base measuring some 620 km in diameter and a flat, caldera-indented summit rising to 27 km above the surrounding landscape, Olympus Mons is one of the largest mountains in the entire Solar System.

Immense crustal tensions have caused extensive rifting in certain areas. Radiating around Tharsis are dozens of large linear rift valleys (termed "fossae"), formed when the Tharsis region uplifted and bulged. The resulting tensions pulled apart the crust in the region, producing faults, rifting and rift valleys. An extensive mass of roughly parallel fossae is to be found in the vicinity of Alba Patera, an immense 450-km-diameter volcanic crater on the northern side of the Tharsis plateau. Along some of these linear valleys, small volcanic craters appear to have been formed, many appearing as chain craters (termed "catenae").

An interconnected system of broad, deep valleys, known collectively as Valles Marineris (named for the Mariner 9 Mars orbiter which imaged it in 1971), cuts into the southeastern side of Tharsis. Almost 5000 km long from east to west, and in places up to 8 km deep, the Marineris system extends almost a quarter of the way around the planet, just south of the equator. The western end of Valles Marineris is a chaotic hive of canyons and gorges known as Noctis Labyrinthus. Dozens of smaller canyons (most of which would dwarf the Grand Canyon in the United States) reach into the sides of the main valley; some have been formed by rifting, others by subsidence. At the eastern end of Valles Marineris, the deep dual furrow named Coprates Chasma feeds into Eos Chasma, which broadens out into an interconnected series of depressed knobbly plains and valleys; some of these valleys appear to have been altered by flowing water, with streamlined terrain around features of pronounced relief. It is an amazing terrain, one of the visual wonders of the Solar System.

Mars' southern hemisphere is occupied by a near-continuous series of dusky tracts, low-albedo areas that are the main features visible through a telescope. Most of these features were given names by Giovanni Schiaparelli in the late 19th century and expanded upon by subsequent observers of Mars. Telescopic observers still use this nomenclature, which was based upon ancient Latin and Mediterranean place names, Biblical and mythological sources. Detailed maps based upon spaceprobe images have been given an extensive nomenclature, all of which has officially replaced the classic telescopic nomenclature. To avoid confusion, however, much of the

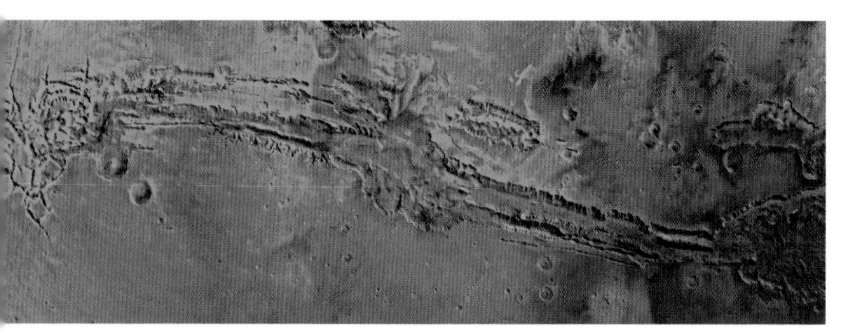

▲ *Mars' Valles Marineris is the Solar System's largest rift valley system – a feature produced when Mars' crust was pulled apart in the remote past.*

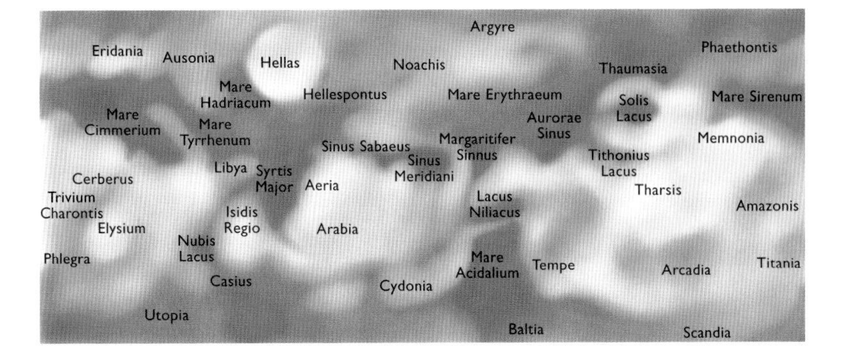

▲ *Albedo map of Mars drawn by the author, with the main features labeled.*

new Martian nomenclature is based around the classical nomenclature of Schiaparelli. For example, the dark area known to telescopic observers as Lunae Lacus (Lake of the Moon) was given the name Lunae Planum (Plain of the Moon) after spaceprobes revealed its true nature.

Although the dark markings on Mars can be considered to be long-lived – though subject to temporary seasonal changes – many of them do not represent substantial topographic features. Instead, most of them reflect the underlying color of the crustal material, and their visibility is subject to seasonal winds which may strip certain areas of its soil covering and obscure others areas with this airborne debris.

First drawn by Christiaan Huygens in 1659, the broad V-shaped wedge of Syrtis Major (Syrtis Planum) is the most prominent of Mars' dark features. On the opposite side of the planet, Solis Lacus is an interesting dark circular feature located within a larger dusky bay, the northern edge of which traces part of Valles Marineris. A broad dark tongue called Mare Acidalium is the most prominent of the far northern hemisphere's dusky tracts. None of these major telescopic features is associated with a topographic formation.

There are some features discernable through the eyepiece that really do mark topographic outlines. For example, the bright areas of Hellas and Argyre mark the circular sites of very large and exceedingly ancient impact basins – Hellas Planitia and Argyre Planitia, 2500 km and 870 km across respectively – whose deep interiors have been filled with wind-deposited soil particles. A number of smaller impact features can also be discerned, notably the 460-km-diameter crater Schiaparelli indenting the northern edge of Sinus Sabaeus (Terra Sabaea) and the identically sized crater Huygens on the northeastern edge of Syrtis Major.

Martian atmospherics

Mars' surface temperature averages −63°C, with minimum temperatures of −140°C (some 50°C colder than the minimum recorded temperature on Earth) and maximum temperatures of a rather pleasant 20°C.

Surprisingly, all the gases making up the Martian atmosphere can also be found in the air that we breathe on the Earth – but in completely different proportions. Carbon dioxide, a minor constituent of the Earth's atmosphere, makes up no less than 95% of the Martian atmosphere. Oxygen, a gas so vital to life on Earth, accounts for only around one part in a thousand of Mars' atmosphere. The remainder is nitrogen, with small amounts of the noble gases. Although an astronaut breathing in a small sample of Martian air would not come to much harm, it contains too little oxygen for human survival. The atmospheric pressure on Mars' surface is also very low, averaging around a hundredth that found on the Earth at sea level.

Even though Mars' atmosphere contains only around one thousandth the amount of water vapor present in the Earth's atmosphere, this is sufficient to cause clouds to form at high altitudes. Orographic clouds form as moisture-laden air is pushed upward over the summits of high mountains. In this manner, striking bright clouds, brilliant enough and large enough to be seen through amateur telescopes, frequently form on the leeward side of Olympus Mons and the other Tharsis volcanoes. Large patches of fog can also form during the

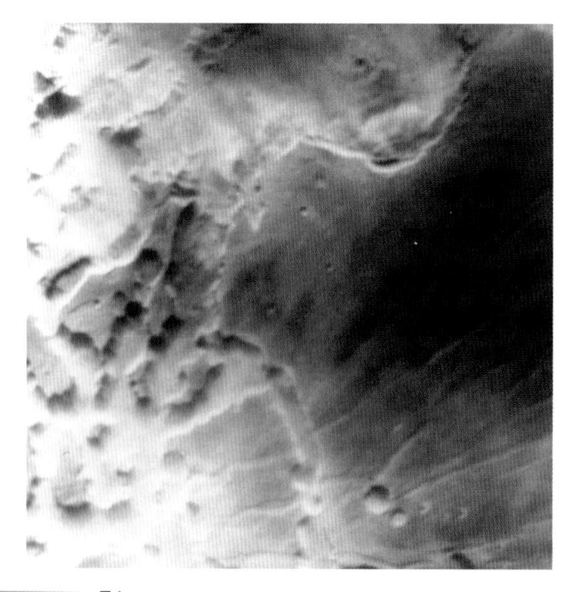

◀ *A Viking orbiter image of early morning fog in Noctis Labyrinthus, a valley system at the western end of Valles Marineris.*

morning in low-lying areas, such as the canyons of Valles Marineris. Water frost lightly coats the surface during the long chill of Martian winter. Mars' low atmospheric pressure prevents water from existing as a liquid on the surface; when the temperature rises above freezing point, water ice sublimates and turns directly into vapor. There is no doubt that in the past Mars' atmospheric pressure was at times high enough to allow water to flow across its surface, forming shallow seas and lakes which were fed by rivers.

As Mars' atmosphere is heated by the Sun, winds are produced, which loft dust high into the air. Large amounts of dust are capable of causing Mars' features to appear muted, occasionally obscuring them from view altogether, on a regional or even global scale, for several weeks on end. Astronomers call these events dust storms, and although they may appear dramatic, on the planet's surface they are far less severe than a typical dust storm in one of Earth's deserts. The fine grains of dust storms reflect around 25% of the sunlight falling on them, so they appear bright in comparison with the planet's darker desert features, which have an albedo of around 10%. Since the amount of solar energy received by Mars varies with season and distance from the Sun, there is a pattern to the Martian dust storms: they tend to be most severe when Mars approaches perihelion, when the planet receives 20% more solar energy than its yearly average.

In mid-2003, as Mars was approaching its perihelic opposition, several regional dust storms were observed, the largest of which spilled out from the Hellas basin and extended southward across Syrtis Major to cover an area of around 600,000 square km in less than a week. Dust storms also brewed up at the end of 2003, threatening to converge and become a global phenomenon, putting into jeopardy the missions of three Mars landers (dust might have settled on each of the craft's solar panels, cutting down the power available for normal operations).

Polar caps

Even a small telescope will reveal Mars' brilliant polar ice caps when the planet is around opposition, and a 150 mm telescope will show the polar caps throughout an entire apparition. Depending on which pole is tilted toward the Earth, the observer is likely to see either the north or south polar cap. Both polar caps are composed of water ice and frozen carbon dioxide (dry ice). Water ice, with its far higher sublimation temperature, forms the small core of each ice cap and is present all year round. An accumulation of carbon dioxide ice – around one meter deep – occurs around each pole during the wintertime, causing the caps to grow in size. As the temperature rises above −120°C during the

Martian summer, the seasonal dry ice caps sublimate, but the residual cap of water ice remains throughout the whole Martian year.

When at its largest during the southern winter, the south polar cap is around 4000 km in diameter – about 20% of the entire surface area of Mars. As the seasonal south polar cap retreats during the summer, its edge begins to show irregularities, and a broad dark tract appears to cleave part of it away from the main body of the cap; the desert regions adjacent to the cap appear to darken during the seasonal cap's retreat. These phenomena are easily visible through a 100 mm telescope at a favorable apparition. As northern winter begins to set in, the seasonal north polar cap grows to about half the area of the seasonal south polar cap at its largest.

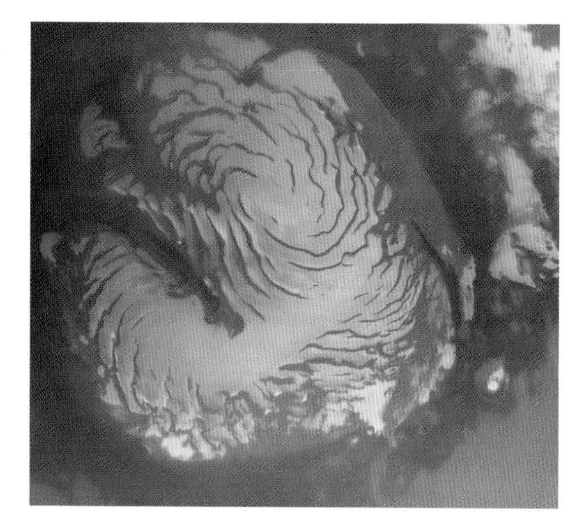

◀ *Mars Global Surveyor imaged the north polar cap in January 2001; it was early summer for the martian northern hemisphere. The intricately structured residual cap is composed of water ice, and remains throughout the summer.*

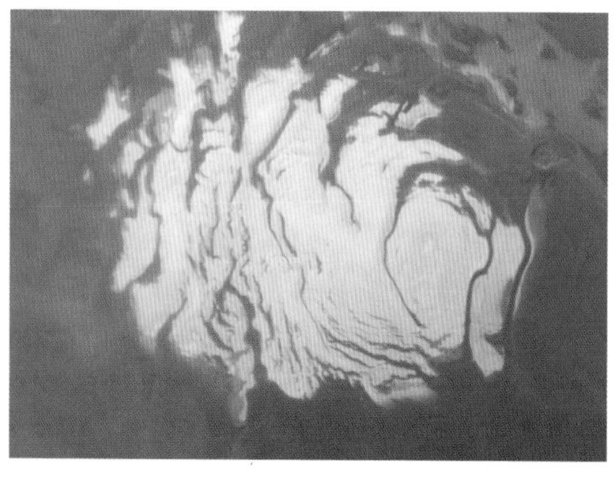

◀ *Mars' south polar cap, imaged by Mars Global Surveyor in April 2000; it was summertime for the planet's southern hemisphere. The ice cap has shrunk to its minimum size, around 420 km across.*

Orbit and rotation

Mars orbits the Sun every 1.88 years (687 days) at an average distance of 227.9 million km. Its orbit is inclined by just 1.9° to the plane of the ecliptic, but it is quite eccentric, with a perihelion distance of 206.6 million km and an aphelion distance of 249.3 million km – a difference of 42.7 million km between its closest and furthest points from the Sun. Mars reaches opposition in the skies of planet Earth every 780 days or so (almost every other year).

Since Mars' equatorial plane is inclined a little more than 25° to the plane of its orbit, Mars experiences seasons, just like the Earth, only they are around twice as long. Southern spring (northern autumn) on Mars lasts 146 days; southern summer (northern winter) 160 days; southern autumn (northern spring) 199 days; and southern winter (northern summer) lasts 182 days on Mars. Seasonal changes are evident through the telescope eyepiece, both in the geometry of the observed features caused by the planet's changing tilt and in their actual physical appearance. Mars' presentation to the terrestrial observer, with regard to its tilt, depends on where Mars lies in its orbit and where the Earth lies in relation to it. The most obvious indicator of the planet's tilt is its polar caps. A 150 mm telescope will usually reveal one or other polar cap at any time during an apparition, when the planet is larger than five arcseconds in apparent diameter.

Rotating on its axis once every 24.6 hours, a day on Mars is just 37 minutes longer than an Earth day. Observational drawings of Mars separated by an hour or so will easily show the planet's rotation. If Mars is observed at about the same time each evening, Mars' rotation will appear to lag behind. The difference amounts to around 9° of Martian longitude per day, and it follows that a complete tour around the planet can be accomplished in around six weeks. Observational drawings made at around the same time on each clear evening over a period of time will clearly show features on the meridian located progressively eastward, giving the illusion that Mars is rotating in a retrograde fashion. For example,

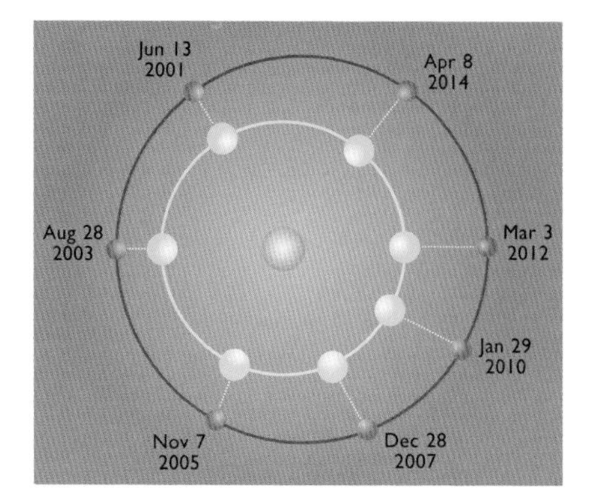

▶ *The orbits of Earth (blue) and Mars (red), to scale. Earth's orbit is nearly circular, but that of Mars is markedly elliptical, producing a wide range of Martian opposition distances and apparent diameters.*

if Sinus Meridiani (on the planet's zero-degree line of longitude) is on the meridian on the first evening, Hellas and Syrtis Major will be on the meridian around a week later; Elysium and Mare Cimmerium will be there around nine days afterward; Tharsis after a further ten days, followed three days later by Solis Lacus, with Mare Acidalium appearing on the meridian around a week later.

Apparitions and oppositions

Oppositions of Mars usually happen every other year, the average interval between oppositions being 780 days (around 50 days later in the year for each successive opposition). Mars is visible for around 18 months in each apparition, during the course of which its apparent diameter grows from a minimum of around 3.5 arcseconds to more than 14 arcseconds (at aphelic oppositions) and 25 arcseconds (at perihelic oppositions). See Table 11 for information on future oppositions.

Mars lies less than 100 million km from the Earth at aphelic oppositions, presenting a disk around 14 arcseconds in apparent diameter which shines at magnitude −1.3. Mars' north pole is always tilted toward the Sun during aphelic oppositions, and features in its blander northern hemisphere are well-presented. At aphelic oppositions, a telescopic magnification of ×120 is required to make Mars appear the same apparent size as the full Moon viewed with the unaided eye.

Perihelic oppositions take place every 15 to 17 years, when Mars approaches the Earth as close as 54.5 million km – around 150 times the distance from the Earth to the Moon. On these occasions, it presents a fully illuminated disk, 25 arcseconds in apparent diameter, and shines at a brilliant magnitude −2.8. Mars' south pole is angled toward the Earth, and the feature-packed southern hemisphere is presented nicely. Although the bright polar ice cap appears prominent, regular observers will note that it has shrunk noticeably in the few months leading up to opposition, as summer temperatures in Mars' southern

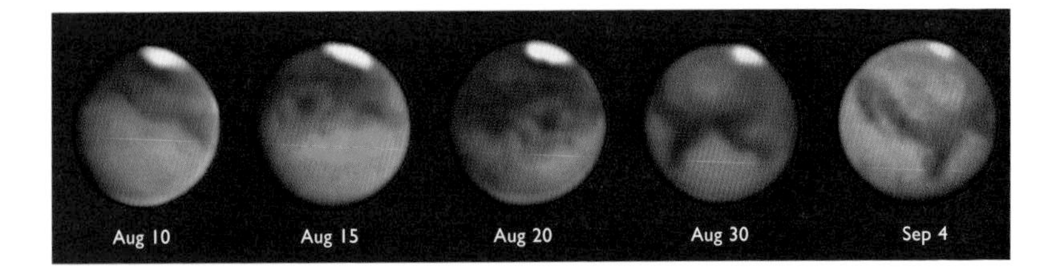

▲ *Mars, imaged by Jamie Cooper around the time of its perihelic opposition in 2003. Mars' rotation period is slightly longer than that of the Earth – note that the position of the central meridian shifts westward on each image.*

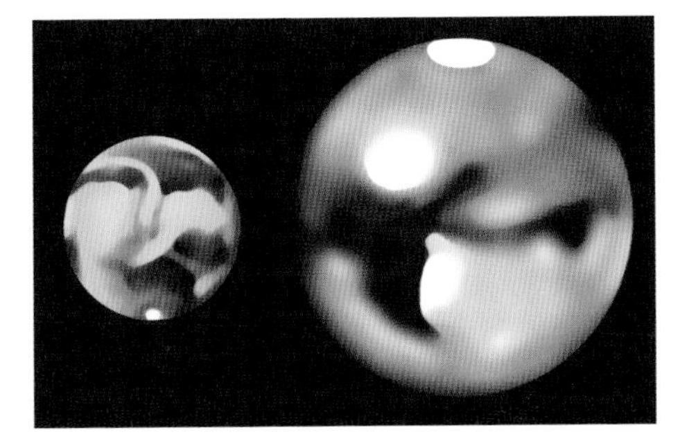

◀ A comparison between the apparent size of Mars at an average aphelic opposition (left, around 14 arcseconds diameter) and at a good perihelic opposition (right, 25 arcseconds diameter).

hemisphere have been rising. It is interesting to note that a good 150 mm telescope will comfortably resolve objects separated by one arcsecond, and at perihelic oppositions, one arcsecond is equivalent to about 280 km on Mars. This is about the same resolution as the naked eye attains when observing the Moon without optical aid. Magnified ×72, the disk of Mars at a perihelic opposition has the same apparent diameter as the full Moon viewed with the unaided eye.

For observers in northern temperate latitudes, perihelic oppositions are hampered by the low altitude of Mars, since perihelion always takes place when the planet is in Aquarius, well south of the celestial equator. The following apparition is always much better suited to observers in the UK and other northerly climes, since Mars exceeds 20 arcseconds in apparent diameter at opposition, while its altitude is much higher, a dozen or more degrees north of the celestial equator in Aries.

Telescopic observation

For several months at the beginning and end of each apparition, Mars presents a small disk, less than 5 arcseconds in diameter, which is rather too tiny to display any intricate detail, even through large amateur telescopes under excellent seeing conditions. For most of an apparition, Mars is not an easy object to observe satisfactorily because of its small apparent diameter. While the broader, darker features are visible through modest-sized telescopes, the fine detail which makes Mars so fascinating to observe can only be discerned when Mars is close to opposition and the atmospheres of the Earth and Mars itself are clear. When Mars exceeds 5 arcseconds in apparent diameter, the larger dusky features and the bright north or south polar cap may be discerned through at least a 150 mm telescope. Most observers consider a Martian disk larger than 10 arcseconds to be the minimum size worthy of serious telescopic scrutiny, neglecting regular observation until Mars is well-established in the night skies, perhaps six months or even longer after its conjunction with the Sun.

A tour of Mars, westward from 0–360°, in four sections

(Classical nomenclature has been used throughout. See also the albedo map on page 75.)

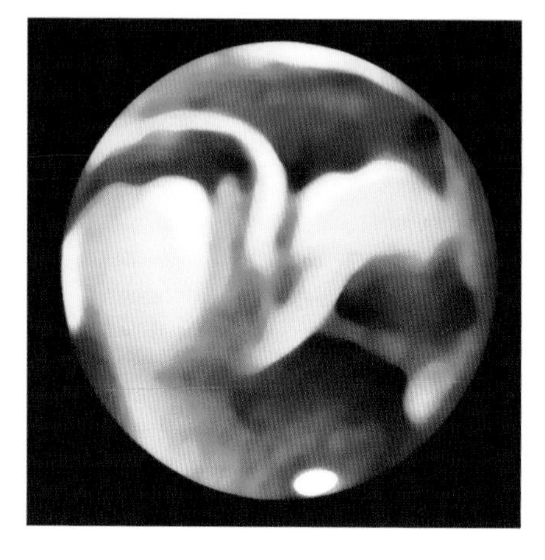

▲ *Mare Acidalium dominates the north and wispy tracts from Sinus Meridiani and Margaritifer Sinus extend toward the dusky Siloe Fons. Observation by the author using a 150 mm Newtonian ×200. May 6, 1999, 23:00 UT. CM 12°.*

Section 1: 0–90°W

During perihelic oppositions, the broad dark region of **Mare Acidalium** in the north – the most clearly defined of Mars' marial areas – appears rather foreshortened near the northern limb. Its true shape and extent can be far more adequately gauged during aphelic oppositions, when Mars' north pole is tilted toward the Sun. At such times, it is an easy feature to discern through a 100 mm telescope. At its largest, Mare Acidalium appears as a broad, dark, mottled tongue, elongated north–south over around 25° of latitude from 60°N to 35°N. Its northwestern margin blends into the area of **Baltia**, with a couple of dark wisps in the form of **Iaxartes** from its far northern edge and **Tanais** extending from its western margin. Mare Acidalium's southern margin is usually rather better defined, often appearing sharpest along its border with the bright circular area of **Tempe** to the southwest. Tempe is centered at around 42°N, 68°W, and unlike some of Mars' brighter circular features it does not appear to be an ancient impact basin. Vague dusky streaks can sometimes be traced across Tempe, along with some brighter patches, including **Nix Tanaica** on the far eastern edge of Tempe, a feature which can appear brilliant.

Extending from eastern Tempe to define the southern edge of Mare Acidalium, a wide, light-toned linear tract called **Achillis Pons** is frequently visible. Immediately to its south, the dark wedge of **Niliacus Lacus** reaches southwest for around 30°, separating southern Tempe from another bright region, **Chryse**. This dark extension, **Nilokeras**, is often observed to be double – it was once cited as the most obvious of the so-called Martian "canals." Nilokeras joins with a small dark patch, Lunae Lacus, at 65°W, 20°N. **Lunae Lacus** is rarely prominent, but several swathes of dark material reach out from it, most notably **Ganges**, which extends south to the Martian equator. Here it meets a small, well-defined dark spot, **Juventae Fons**, usually visible through a 150 mm telescope at a favorable apparition. In favorable circumstances, the feature appears to be joined by an exceedingly narrow causeway to the delicate frog's hand-shaped **Aurorae Sinus**, centered at around 12°S, 55°W, from which spreads a dusky web of tracts toward the north and west. Of these, **Agathodaemon** curves for nearly 15° of longitude across to **Tithonius Lacus**, a dark spot usually somewhat larger and easier to discern than Juventae Fons, lying some 10° to its west. Agathodaemon and Tithonius Lacus mark the position of much of the magnificent Valles Marineris rift valley system.

Southwest of Aurorae Sinus an occasionally broad dark tract called **Nectar** runs west to join one of Mars' most prominent features, **Solis Lacus**. Centered at 25°S, 90°W, Solis Lacus is a roughly circular patch which varies in extent and intensity from one apparition to the next. At its largest, it covers more than 20° in longitude. It sits just south of the central position within a brighter area bounded in the south by the dark curving tract of **Bosporos**, usually separated from it by the light-toned **Thaumasia**. Sometimes the southern edge of Solis Lacus blends into a darker than usual Thaumasia, making its outline difficult to trace. It is always pretty clearly defined in the north. Occasionally, Solis Lacus appears as a mottled collection of dark, ill-defined patches, but sometimes it can appear as a striking single spot, justifying its unofficial name of the "Eye of Mars."

South of Aurorae Sinus, stretching a quarter of the way around Mars from around 0°W to 90°W, between latitudes 20°S and 40°S, is **Mare Erythraeum**, which appears as a collection of dark patches. In the south, Mare Erythraeum fades and embraces the large and usually light-toned, clearly defined circular patch of **Argyre**, an ancient impact basin some 870 km in diameter, centered at 49°S, 38°W.

Mare Erythraeum darkens in the north at around 20°W, and links with **Margaritifer Sinus**, a dark, narrow V-shaped feature

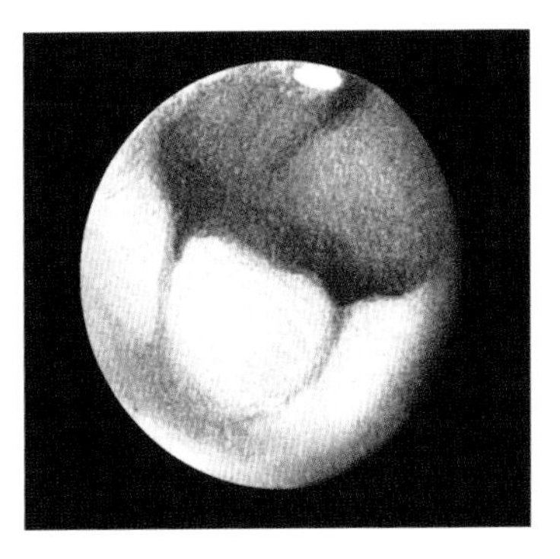

▲ *Mare Erythraeum is prominent in the south, with wisps from Sinus Sabaeus and Aurorae Sinus leading north to a muted Mare Acidalium. The south polar cap is small. Observation by the author using a 100 mm refractor ×200. November 3, 1988. 20:50 UT. CM 210°.*

which extends and narrows north of the equator, curving slightly eastward around the margin of Chryse. Margaritifer Sinus usually shows a paler tone than Aurorae Sinus to its west, but its northern tip sometimes appears very pointed and extends along a tract called **Oxus** to the dusky node of **Siloe Fons**, southeast of Mare Acidalium.

Section 2: 90–180°W

Tharsis, a large volcanic plateau, straddles the equator between around 80°W and 120°W. Despite the gigantic nature of these volcanoes, it is difficult to see them directly through backyard telescopes. **Nix Olympica** (Olympus Mons), located at 21°N, 127°W, is often visible because of the bright orographic clouds that form near its summit; long before its mountainous nature was known, the feature merited its highly appropriate name – truly an Olympic mountain, fit for the gods! Under excellent circumstances, three faint dusky spots can be seen near Nix Olympica – **Ascraeus Lacus**, **Pavonis Lacus** and **Arsia Silva** – the large sister volcanoes of Olympus, which form a northeast–southwest line immediately between Nix Olympica and Solis Lacus.

A host of other faint features may be glimpsed in the region, including **Nodus Gordii** on the equator at 135°W and various ill-defined dusky zones within **Amazonis** and **Arcadia**, both light-colored zones north of the equator. The far northern latitudes are considerably less intensely tinted than the southern hemisphere, but to the west of Arcadia a patchwork of dusky features – including **Castorius Lacus**, **Propontis**, **Herculis Pons** and **Stymphalius Lacus** – bubbles southward. South of the equator, the dark crescent of **Mare Sirenum** borders the brighter region of **Memnonia**. When Mare Sirenum is on the central meridian, it is likely to be the darkest feature visible. **Fusca Depressio**, a small lobe on its northeastern border, is usually the most intensely dark part of the mare. South of Mare Sirenum are a number of broad, light-toned regions, including **Phaethontis** and **Electris**; further south, **Thyle** extends toward the south polar cap.

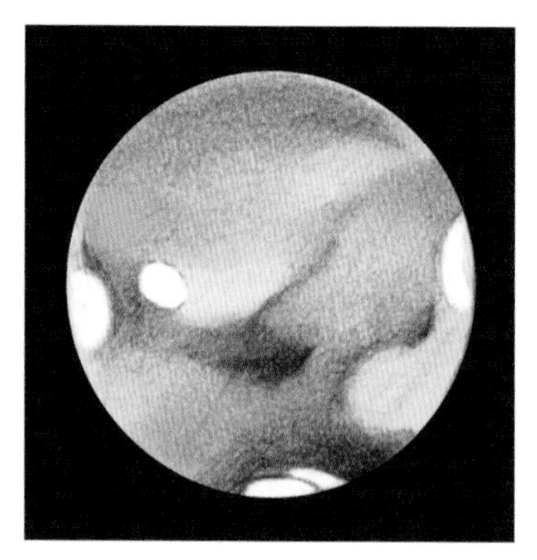

▲ *Bright clouds over Tharsis, including an isolated cloud over the giant volcano Olympus Mons, are approaching the western limb. Observation by the author using a 225 mm Newtonian ×250. March 31, 1997, 02:30 UT. CM 168°.*

Section 3: 180–270°W

Extending from Propontis, spanning 20° of latitude, the ill-defined tract of **Phlegra** meets **Trivium Charontis**, a dark feature centered at 20°N, 197°W. **Cerberus**, a prominent wide, dark tract, extends southwest of Trivium Charontis, skirting the southern edge of the bright circular **Elysium**, a volcanic plateau centered at 23°N, 215°W. These features are often easily visible through a 100 mm telescope. The western edge of Elysium is not so clearly delineated, but faint markings can often be seen in the regions of **Aetheria** and **Aethiopis**. Nearby, at 35°N, 257°W, **Nubis Lacus** forms a sometimes clearly defined dusky node, into which spiral several ill-defined tracts, such as **Casius** emanating south from **Utopia**, and **Thoth-Nepenthes** emanating from **Lacus Moeris** just east of **Syrtis Major** (see below).

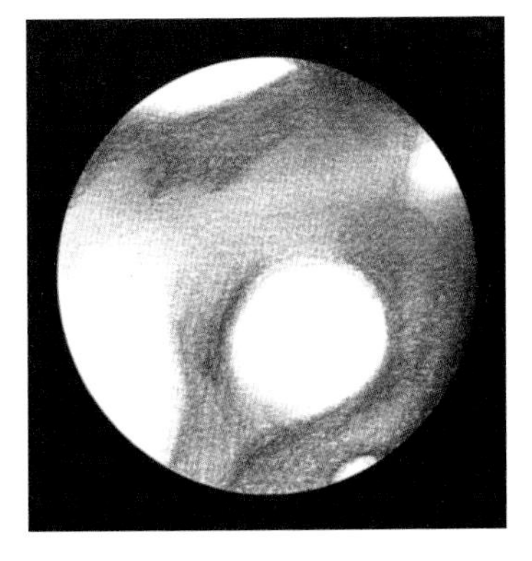

▲ *Elysium appears as a bright area on the central meridian, bordered on the west by Trivium Charontis. Mare Erythraeum is a dark region in the south, bordered near the limb by a bright Eridania. The north polar cap is small but bright. A brilliant feature on the terminator in the east is a possible cloud formation in the region of Libya. Observation by the author using a 250 mm Newtonian ×240. May 21, 1999, 21:20 UT. CM 215°.*

Just south of the equator, Cerberus joins with the northwestern reaches of **Mare Cimmerium**, a broad dusky arc stretching almost 70° in longitude from Mare Sirenum to 245°W. Mare Cimmerium is an easy object through small telescopes. It is darkest and most clearly defined along its northern margin, and displays considerable detail through a 200 mm telescope under good seeing conditions. **Eridania** is occasionally easy to discern as a light patch south of Mare Cimmerium, and to its south **Mare Chronium** is marked by a hazy streak.

Hesperia, a light zone, separates Mare Cimmerium from **Mare Tyrrhenum**. Sometimes the division between the maria is not too distinct, but Mare Tyrrhenum is usually the darker of the two, although it can display considerable patchiness. The northern edge of Mare Tyrrhenum extends into a narrow pointed feature called **Syrtis Minor**, whose northern tip just touches the equator. **Ausonia**, a region south of Mare Tyrrhenum, sometimes appears rather bright, but at other times is indistinct.

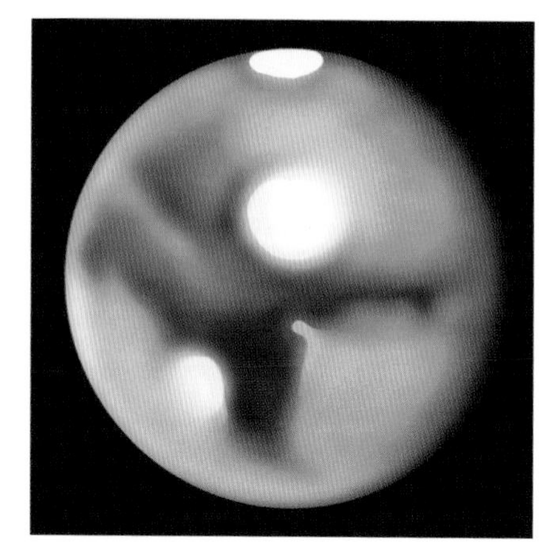

▲ *Syrtis Major is prominent, with a bright Hellas and Libya. The small notch at the junction of Syrtis Major and Sinus Sabaeus is formed by the crater Huygens. Observation by the author using a 300 mm Newtonian ×250. October 30, 1988. 00:10 UT. CM 303°.*

Section 4: 270–360°W

Mare Tyrrhenum flows northwest toward the equator, where it blends with the southeastern corner of **Syrtis Major**, a very prominent broad dark wedge visible in the smallest telescopes at opposition. Syrtis Major protrudes northward across the equator to around 20°N latitude. Its western edge, bordering the bright area of **Aeria**, is often the darkest and most clearly defined part of the feature; its eastern edge, bordering **Libya** and **Isidis Regio**, displays marked seasonal variations. A clearly defined bright patch within Isidis Regio, just north of the protruding **Lacus Moeris**, is sometimes striking, even through small telescopes. Syrtis Major often appears distinctly mottled, and its northern section occasionally appears separated from the rest by a narrow bright lane called **Arena**. During some apparitions Syrtis Major's northern end is sharp and pointed, at other apparitions it appears somewhat blunted and angular. Its northern tip sometimes curves and extends further northward along a prominent curved streak called **Nilosyrtis**, where it meets the dusky tract of **Boreosyrtis**, at around 50°N. At some apparitions, the northwestern edge of Syrtis Major extends to a dark point, from which a fine dark streak called **Astusapes** is sometimes visible.

Arabia, a wide light-toned desert area, blends into the usually brighter region of Aeria. Its northern edge, at around 40°N, is defined by the dark tracts of **Protonilus** and **Ismenius Lacus**. To their southwest, the whole desert region of **Eden** spans more than 40° latitude, but displays only very subtle tonal variations.

Syrtis Major's southwestern corner is often cut through by **Oenotria**, a light-toned strip which borders the dusky patches of **Iapygia** and **Deltoton Sinus**. A distinct semicircular notch in Iapygia at 14°S, 304°W, is sometimes easily visible through a 100 mm telescope. This bay of eastern Aeria actually marks the position of **Huygens**, an asteroid impact crater some 456 km in diameter. South of Aeria, extending west of Iapygia for some 40° of longitude, is the prominent dark tract of **Sinus Sabaeus**. More often at its darkest and most clearly delineated along its northern edge, this feature joins with **Sinus Meridiani**, the famous "forked bay," whose dual prongs extend northward across the equator, straddling the 0° line of longitude. A dark peninsula called **Sigeus Portus** is often visible midway along the northern edge of Sinus Sabaeus; a clearly discernable bay between this feature and Sinus Meridiani marks the position of another large Martian impact crater, the 461-km-diameter **Schiaparelli**. Adjoining the southern edge of Sinus Sabaeus is the variable light-toned tract of **Deucalionis Regio**, which flows around the western margin of Sinus Sabaeus into the bright desert region of **Thymiamata**. Another dusky tract, **Pandorae Fretum**, lies along the southern edge of Deucalionis Regio, and curves northward into the dark V-shaped **Margaritifer Sinus**.

Often the most clearly defined bright area on Mars, **Hellas** lies around 40° due south of Syrtis Major. Its margin is bordered by the dusky **Mare Hadriacum** in the northeast, **Mare Ionium** in the northwest, **Mare Amphitrites** in the southwest and **Mare Australe** in the south. A dark tract called **Hellespontus** curves southwest from Mare Ionium, defining the southern border of the light-toned desert region of **Noachis**.

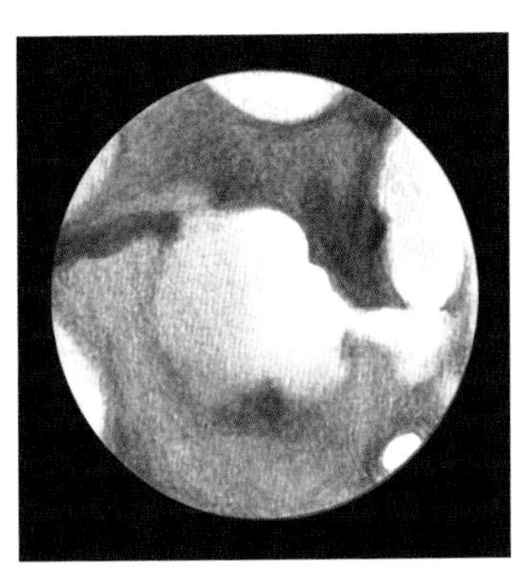

▲ *Syrtis Major, very dark and prominent, is approaching the central meridian. To its south is a very bright and clearly defined Hellas. Detail around Neith and Nubis Lacus is evident. Elysium creates a narrow bright sliver near the terminator. Observation by the author using a 150 mm Newtonian ×200. April 11, 1999, 01:30 UT. CM 276°.*

Recording Mars

Observational drawings of Mars are usually made on 50-mm-diameter circular blanks, but the ALPO Mars Section uses a 42-mm blank (the reasoning: Mars is 4200 miles in diameter). Before observing, it is advisable to indicate the planet's phase and the orientation of its axis on the outline blank, rather than having to guess this at the eyepiece. Tabular information given in various detailed Martian ephemerides will allow an accurate blank to be constructed. To do this, the *BAA Handbook* gives the following relevant information: P = position angle of the north pole, measured east from the north point of the disk; Q = position angle of the point of greatest defect of illumination (that is, the point where the phase is at its greatest width), measured east from the north point of the disk; Phase; Tilt = tilt of Mars' north pole toward or away from the Earth. This information is given in intervals of 10 days, so some interpolation of the data is usually required. Numerous astronomical computer programs will display a graphic showing an accurate grid superimposed upon an image of Mars.

A prior knowledge of the planet's central meridian during the observing session is also a considerable advantage. Astronomical ephemerides, including the *BAA Handbook*, give data on the central meridian's position for 0h each day, which can then be added to or subtracted from according to the time of the observation (see below, Central meridian transit timings, and refer to Table 10). Many observers avoid the mathematics by consulting a suitable computer program to display a graphic representation of the hemisphere presented during the observing session. However, it is important to bear in mind that the shapes and intensities of the features shown on the computer monitor are likely to differ in many ways from the actual view through the eyepiece, so it must be considered as a guide only.

An obvious starting point for an observational drawing is the planet's bright polar cap. Next, the outlines of the more prominent darker

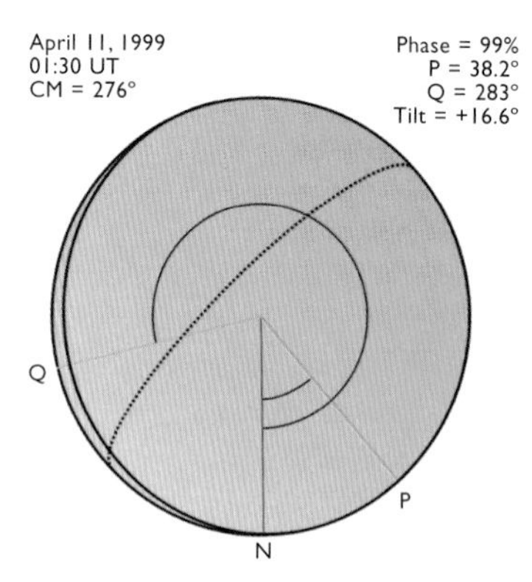

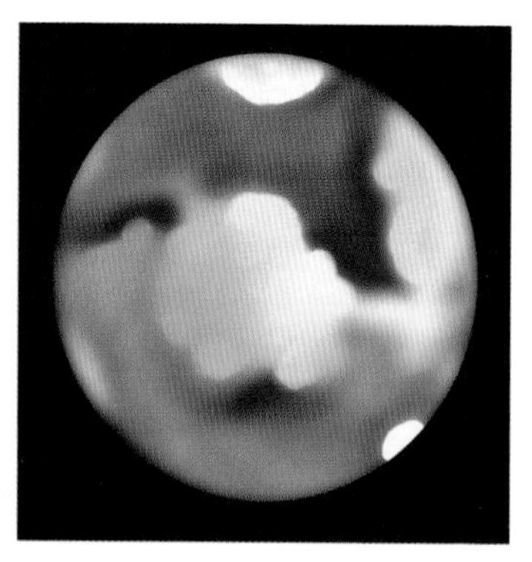

▲ *Construction of a Mars observing blank with the correct phase and orientation. The following are noted: P = position angle of the north pole, measured east from the north point of the disk; Q = position angle of the point of greatest defect of illumination (the point where the phase is at its greatest width), measured east from the north point of the disk; phase; tilt = tilt of Mars' north pole toward or away from the Earth. This example is an observation made at 01:30 UT on April 11, 1999, by the author using a 150 mm Newtonian ×200.*

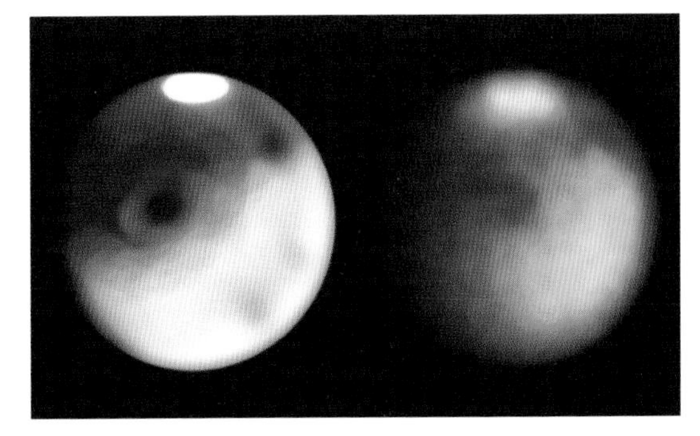

▶ *A comparison between an observational drawing (left) and a CCD image (right) taken minutes apart by the author using the same instrument. Solis Lacus is near the central meridian. Observation made on July 10, 2003, with a 150 mm refractor.*

features can be lightly sketched in, along with any particularly well-defined bright features. Fine detail and tonal shading should be left until last, once the outlines of the main features have been decided upon and drawn boldly. No drawing of Mars, however detailed, is likely to convey a visual impression fully, so many observers make written notes to accompany their sketches (in addition, of course, to the vital observing data which ought to accompany every observational drawing); it is useful, for example, to mention any obscure or uncertain features that may have been observed but not adequately depicted.

Because Mars' surface features may be subject to temporary seasonal changes, or changes taking place over a number of apparitions, some indication of the apparent intensities of features is a useful way of monitoring these variations. Mars observers often use an intensity estimates scale of 0 to 10, with 0 representing the brilliant white polar cap, and 10 representing the black night sky. An intensity estimate of 2 might usually be given to Mars' bright desert regions, and particularly dusky features may appear as dark as 7.

Filter work

Integrated light observations reveal a great deal of Mars' surface features and its atmospheric phenomena, but the use of filters enables much more intensive scrutiny of the red planet's markings and atmosphere. Importantly, filter work requires the image produced in a given light to be reasonably bright – it's pointless to attempt to observe using a filter which gives a very dim image. A red, orange or yellow filter will make the dark surface features easier to see and will improve the definition of less distinct markings in brighter desert regions. Yellow-colored dust storms blow up on Mars from time to time, and they appear brighter through these filters, especially in red light. The most commonly used filter types are red (Wratten 25), orange (Wratten 21) and yellow (Wratten 15).

Atmospheric limb brightening, haze and orographic clouds are revealed using the blue (Wratten 44A and 80A), green (Wratten 58) and blue-violet (Wratten 47) filters, which maintain the relative brilliance of the polar regions, but considerably mute the appearance of the surface features. A phenomenon known as the "blue clearing" occasionally takes place in the Martian atmosphere, when, using a blue filter (observers use the Wratten 47 as a standard for this), surface features become easier to see, perhaps over a period of several days; its precise cause is poorly understood. At least a 200 mm telescope is required to use the Wratten 47 filter effectively, because of its low light transmission.

To assess the state of the Martian atmosphere and the intensity of the blue clearing, the following scale is used:

0	No dark surface features are discernable
1	Some dark surface features are vaguely visible
2	Dark surface features are easy to see
3	Surface features are almost as clearly defined as in integrated light (very rare)

Central meridian transit timings

Central meridian timings are a valuable means of mapping a planet's features. Making accurate timings of the passage of Martian features across the planet's central meridian is hampered by the fact that Mars often displays a phase, making it difficult to judge the position of the central meridian.

When either of Mars' poles is tilted strongly toward the Earth, features appear to make a strongly curved path across the disk as the planet rotates. Prior to opposition, when the planet's western limb is fully illuminated, a prominent feature will appear near the western limb and proceed eastward to transit the planet's central meridian (the imaginary line running down the center of a planet's disk, joining both poles) in around five and a half hours; this feature will disappear into the planet's evening terminator in the east in less time than it took to reach the meridian – the exact time period depends on the planet's phase. At around opposition, the planet is fully illuminated and both east and west limbs are clearly defined. After opposition, the western limb is encroached upon by the morning terminator, and features will reach the central meridian from the terminator in a far shorter period than they take to move from the central meridian to disappear near the fully illuminated eastern limb.

Mars rotates on its axis more than twice as slowly as Jupiter, as can be seen from Table 10, but it is possible to estimate a Martian feature's central meridian passage within a few minutes of accuracy. Although one point along Mars' central meridian can usually be judged from the position of whichever bright polar cap is presented, it is always helpful to consult a detailed planetary ephemeris or a computer program to calculate the precise orientation of the planet's axis and the angle of its central meridian.

While central meridian transit timings of regular Martian features may not be of immense scientific value, they are enjoyable to conduct, and will enable the observer to draw up an independent map of the planet's features based on calculations of the planet's longitude given in astronomical ephemerides and computer programs. Transit timings can be of scientific value in pinpointing the location of any unusual Martian features, such as bright orographic clouds or localized dust storms, or any unusual transient phenomena. Importantly, the observer ought never to assume that his or her observations are pointless because someone else with a bigger telescope and better equipment is doing the observing at the same moment. Many major planetary discoveries have

TABLE 10 CHANGE OF LONGITUDE IN INTERVALS OF MEAN TIME

h	°	h	°
1	14.6	6	87.7
2	29.2	7	102.3
3	43.9	8	117.0
4	58.5	9	131.6
5	73.1	10	146.2

min	°	min	°	min	°
10	2.4	1	0.2	6	1.5
20	4.9	2	0.5	7	1.7
30	7.3	3	0.7	8	1.9
40	9.7	4	1.0	9	2.2
50	12.2	5	1.2	10	2.4

TABLE 11 OPPOSITIONS OF MARS 2005–2018

Date	RA	Dec	Size (arcseconds)
2005 Nov 7	02h 51m	+15° 53′	19.8″
2007 Dec 28	06h 12m	+26° 46′	15.5″
2010 Jan 29	08h 54m	+22° 09′	14.0″
2012 Mar 3	11h 52m	+10° 17′	14.0″
2014 Apr 8	13h 14m	−05° 08′	15.1″
2016 May 22	15h 58m	−21° 39′	18.4″
2018 Jul 27	20h 33m	−25° 30′	24.1″

been made by keen amateurs prepared to observe and make records of Solar System objects during periods when most other observers are likely to be neglecting that particular object near the beginning or end of its apparition.

Example:
On August 28, 2003, the northern tip of Syrtis Major was noted to cross the central meridian of Mars at 22:18 UT. The ephemeris gives a figure of longitude 320.5° for Mars' central meridian at 00h on August 29. The difference between 22:18 and 00h is 1h 42min. Converting this difference to longitude, 1h = 14.6° + 40min = 9.7° + 2min = 0.5°, gives a total of 24.8°. Deducting 24.8° from 320.5° gives a longitude of 295.7° for the northern tip of Syrtis Major, as observed on that evening. See Table 10 to calculate the drift of the CM longitude over a given time period.

The Martian moons

Phobos and Deimos, the two satellites of Mars, are potato-shaped city-sized lumps of rock orbiting almost directly above the Martian equator. Phobos orbits at a distance of a mere 2583 km above the surface of Mars in a period of just 7h 39m – the only satellite to orbit its primary faster than the primary revolves. Deimos orbits 16,642 km above the Martian surface in a period of 1d 6h 18m.

Both moons are made of very dark material, and reflect only around 6% of the sunlight falling upon them. Phobos measures about 27 km long by 19 km wide, while Deimos is around half its size. Both moons display impact cratering. Phobos is dominated by Stickney, a deep crater measuring around one-third of Phobos' length. Much of Phobos' surface is striated with parallel grooves, composed of interconnected crater chains with diameters ranging from a few tens of meters to several hundreds of meters. These odd-looking features were probably formed by a multitude of simultaneous gaseous eruptions following the asteroidal impact which created Stickney. Water ice and other volatiles in Phobos' interior would have superheated, causing violent explosions to appear across the satellite's surface. Deimos' surface is covered with a thick blanket of soil and rock, which makes it appear somewhat smoother than Phobos.

Discovered in 1877 by Asaph Hall at the eyepiece of the 675 mm refractor of the US Naval Observatory, Phobos and Deimos are exceedingly faint objects. They require a sizable telescope (at least 300 mm), good Martian opposition circumstances, a steady atmosphere and a keen eye to observe visually. It is important to be aware of the precise positions of the satellites at the time of the observation:

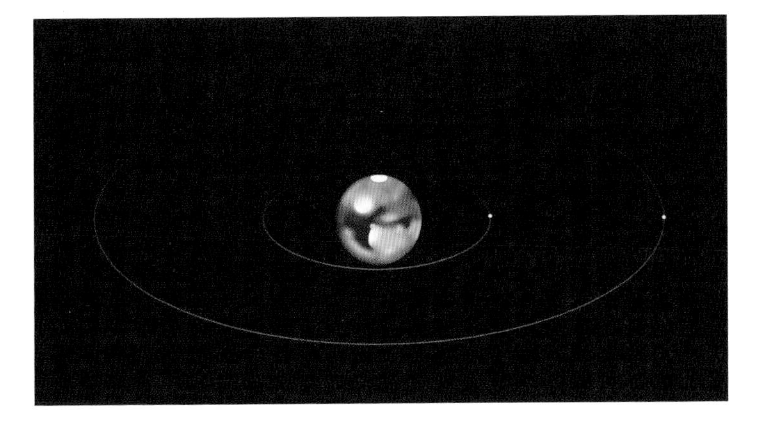

▲ *The orbits of the martian moons Phobos (inner) and Deimos (outer) drawn to scale by the author. Under the right conditions it is possible to discern these satellites visually using a 300 mm telescope.*

a good astronomical computer program or a website such as the NASA/JPL Solar System Simulator at http://space.jpl.nasa.gov/ can be used to find out this information. Both moons move around Mars at such a rapid speed that there is only a small window of opportunity to view them at their maximum elongations from the planet, when they are as far away as possible from the red planet's glare. Deimos is the least difficult to discern when at a maximum elongation from Mars – around 3.5 Martian diameters from the planet's center and shining at a feeble magnitude 12.9 at a close perihelic opposition. With Mars in the field of view of an ordinary eyepiece, its glare renders the satellite virtually impossible to see, so the disk of Mars needs to be hidden from direct view. Hiding Mars will eliminate most of the intrusive light, but some residual glare will always be visible due to diffraction and irradiation, even if Mars is just out of the field of view. It is possible to fabricate a makeshift occulting mask from a small piece of tin foil inserted near the eyepiece's focal plane, which will enable Mars to be positioned within the field at the same time that the satellite search is taking place.

6 · MINOR PLANETS

Hundreds of thousands of minor planets have been identified since Giuseppe Piazzi discovered the first one, Ceres, on January 1, 1801, and the search continues. It is still possible for a genuine asteroid discovery to be made by any amateur who takes the trouble to carefully examine photographic or CCD images – the images don't necessarily have to be his or her own!

More than one million asteroids are estimated to orbit within the main asteroid belt between 2.5 and 4.5 AU from the Sun (1 AU, or Astronomical Unit, is the average distance between the Earth and the Sun), orbiting in a variety of inclinations to the plane of the ecliptic. In addition to the main asteroid belt, asteroids have been found to occupy a range of other orbits within the Solar System. As of the beginning of 2005, more than 277,000 minor planets have had their orbits calculated; of these, nearly 100,000 have been designated official numbers, but only around 12,000 of these have been given names by their discoverers. Of these, more than a thousand are bright enough at opposition for them to be seen through the telescope eyepiece.

Near-Earth asteroids

A group of minor planets known as the Amor asteroids (named for asteroid 1221 Amor) cross from beyond the orbit of Mars to approach the Earth's orbit; more than a thousand of this group have been identified. Apollo asteroids (named for 1862 Apollo) actually cross the orbit

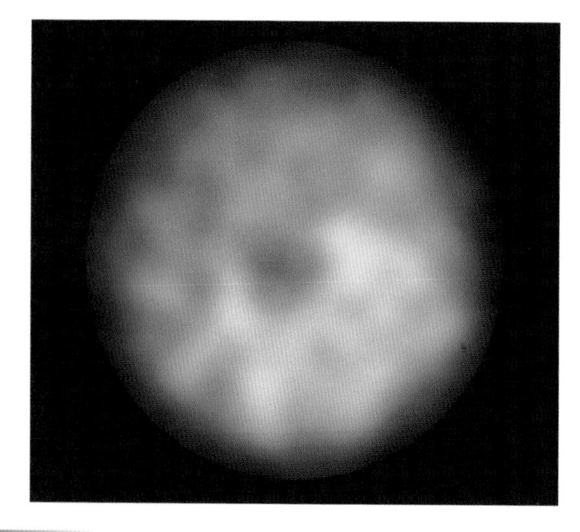

◀ Ceres, imaged by the Hubble Space Telescope in June 1995. The small dark spot near the center of the asteroid's circular disk has been named Piazzi after Ceres' discoverer, Giuseppe Piazzi.

of the Earth – hence their rather terrifying nickname of the "Earth-grazers." More than 1100 Apollo asteroids are known, many of which pose a possible impact threat to the Earth in the far future. Orbiting entirely within the orbit of the Earth, more than 180 Aten asteroids (named for 2062 Aten) are known. The brightest examples of all of these asteroid groups are sometimes visible through the eyepiece of an average backyard telescope.

Many asteroids within the Earth-orbit-crossing Apollo group are termed "potentially hazardous asteroids" (PHAs). Since the definition of a PHA is that its orbit must approach within 7.5 million km (0.05 AU) of the Earth's orbit, and the asteroid must be large enough to constitute a threat to humanity, a number of Amor and Aten asteroids actually fall within the PHA category.

An asteroid venturing very close to the Earth appears to move across the sky very rapidly, and its motion may be visible in real time through the telescope eyepiece. Discoveries of near-Earth objects are often made just a few weeks ahead of a close encounter, giving astronomers only a short time to prepare for the event. Close approaching asteroids sometimes brighten enough for them to be seen through large binoculars and telescopes. Finder charts and ephemerides for close approaches of near-Earth asteroids are occasionally published in astronomy magazines, but often the lead-up time to encounter following discovery is too short to be published. Many astronomical societies (such as the Society for Popular Astronomy in the UK) produce electronic news alerts which are emailed out soon after important discoveries are made – particularly concerning those objects that can be observed by the amateur astronomer. A number of astronomical websites are frequently updated with the latest news and contain ephemerides and finder charts for close encounters with recently discovered asteroids.

One particularly notable example of a near-Earth asteroid which will become visible to the unaided eye during a close encounter is asteroid 2004 MN4, which will zip past the Earth on April 13, 2029, brightening to magnitude 3.3, and will be visible from Europe, Africa and western Asia. Its small disk and some surface features ought to be resolvable at a high telescopic magnification. A series of similar close approaches by 2004 MN4 are set for 2035, 2036 and 2037, but its precise orbit following 2029 cannot be calculated in detail in advance because of uncertainties introduced by the strong gravitational perturbation by the Earth.

Far-out space rocks

In the far reaches of the Solar System, the 130-plus known Centaur asteroids orbit at distances between 5.5 and 25 AU from the Sun. An unusual class of asteroids called the Trojans orbit at the same distance from the Sun and along the same orbital plane as Mars, Jupiter and Saturn, clustered around the Lagrangian points – regions 60° in advance of and following these planets where gravitational resonance allows them to maintain stable orbits. Finally, at the fringes of the Solar System, Kuiper Belt Objects are a group of more than 700 currently known asteroids and inactive comets orbiting beyond Neptune. None of these objects are actually visible through the eyepiece of an amateur telescope, but amateurs have succeeded in imaging many of them using CCD cameras.

Main-belt asteroids

Of those minor planets orbiting in the main asteroid belt between Mars and Jupiter, Ceres is the largest, with a diameter of around 1000 km. There are 26 asteroids larger than 200 km, including Pallas (610 km), Juno (240 km) and Vesta (540 km). It was once suggested that the asteroids may represent the scattered fragments of a planet which somehow exploded. Studies of meteorites, which are chips off larger asteroidal bodies, show that some have come from an ancient body large enough to have been molten inside and undergone differentiation, forming a rocky crust, mantle and iron-rich core. These ancient planets formed alongside the other planets, but experienced frequent mutual collisions, breaking them up. They were not huge objects to begin with – it has been estimated that if all the known asteroids were gathered together they would form an object only around half the size of our own Moon.

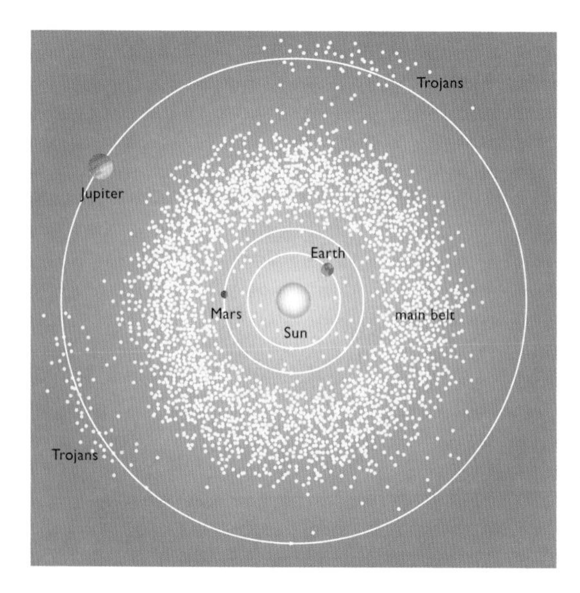

◀ The Solar System's main asteroid belt occupies a broad zone of space between Mars and Jupiter. Also shown here are the Trojan asteroid groups, which lead and trail Jupiter by 60°.

TABLE 12 LARGEST MINOR PLANETS IN THE MAIN ASTEROID BELT

Minor planet	Diameter (km)	Mean distance from Sun (AU)
1 Ceres	1000	2.766
2 Pallas	610	2.773
4 Vesta	540	2.361
10 Hygiea	450	3.136
31 Euphrosyne	370	3.148
704 Interamnia	350	3.067
511 Davida	325	3.170
65 Cybele	310	3.437
52 Europa	290	3.099
624 Hektor	285	5.203
451 Patientia	275	3.060
15 Eunomia	270	2.644
16 Psyche	250	2.919
48 Doris	250	3.109
92 Undina	250	3.189
324 Bamberga	245	2.682
3 Juno	240	2.667
24 Themis	235	3.129
95 Arethusa	230	3.073

In recent years, our knowledge of the minor planets has increased dramatically. Large optical telescopes based on the ground and in Earth orbit have used new technology and techniques to produce images of the surfaces of a number of large asteroids. Radio telescopes have also been used to image the surfaces of some asteroids as they have approached close to the Earth. Asteroids come in all sorts of shapes and sizes; their gravity is so small that the solid rock from which they are made resists the pull of gravity. They have a generally rounded shape, rather than being sharp and angular, because their surfaces have been subjected to eons of relentless bombardment from smaller meteoroids, which have sandblasted and eroded the surface, producing a pulverized layer of topsoil. Larger impacts have produced craters; each asteroid so far imaged has been covered with impact craters.

Observing main-belt asteroids

The four biggest main-belt asteroids – Ceres, Vesta, Pallas and Juno – are well-suited to observation through binoculars and small telescopes for months on either side of their opposition (see Tables 14–17). Although small instruments will show asteroids as tiny points of light (except under the very rare circumstances of a close approach, see below), these objects are nevertheless fascinating to observe and follow across the sky. A number of the brightest asteroids, including Ceres, Vesta, Pallas and Juno, can even be followed through a pair of 7 × 50 binoculars for a period of time on either side of opposition; indeed, under good conditions, Vesta can be seen with a keen unaided eye when it is around opposition, shining at an average magnitude of 6.5, brightening to magnitude 5.2 (slightly brighter than Uranus) in some years.

An 80 mm rich-field refractor on an altazimuth mount or a pair of good, mounted binoculars (at least 7 × 50s) are perfect instruments for hunting down the big four asteroids. Detailed finder charts for the big four and other notable asteroids frequently appear in the pages of popular astronomical magazines. Asteroid data and ephemerides are published in the *BAA Journal* and other celestial almanacs, and there is no shortage of websites giving this information, among them the ALPO and BAA. Many planetarium computer programs show the paths of the brighter asteroids to a high degree of accuracy, and some have the capacity for adding the orbital elements of newly discovered objects, such as those which might make a close flyby of the Earth, for example.

Since asteroids can appear exceedingly faint, it's best to attempt to search for those whose predicted brightness is at least a magnitude greater than the faintest stars visible through your instrument. Perform a pre-hunt check on a star pattern with a range of known magnitudes

TABLE 13 BRIGHTEST MINOR PLANETS

Minor planet	Mean opposition magnitude	Minor planet	Mean opposition magnitude
4 Vesta	6.5	44 Nysa	9.9
1 Ceres	7.4	29 Amphitrite	10.0
7 Iris	7.8	19 Fortuna	10.1
2 Pallas	8.0	23 Thalia	10.1
6 Hebe	8.3	10 Hygeia	10.2
15 Eunomia	8.5	42 Isis	10.2
3 Juno	8.7	43 Ariadne	10.2
8 Flora	8.7	63 Ausonia	10.2
18 Melpomene	8.9	44 Nysa	9.9
9 Metis	9.1	29 Amphitrite	10.0
20 Massalia	9.2	19 Fortuna	10.1
5 Astraea	9.8	23 Thalia	10.1
14 Irene	9.8	10 Hygeia	10.2
11 Parthenope	9.9	42 Isis	10.2
12 Victoria	9.9	43 Ariadne	10.2
16 Psyche	9.9	63 Ausonia	10.2
27 Euterpe	9.9		

(the area around Polaris or Crux, for example) and determine the faintest star visible. It would be frustrating to try to search for an asteroid around your limiting magnitude, since a faint background star might easily be mistaken for your asteroidal quarry.

When using a computer program to display the positions of asteroids and to generate observing charts, set it up so that it shows only the stars brighter than your limiting magnitude. This eliminates a great deal of confusion, both on screen and on the printout, since only those stars which can actually be seen through your instrument will be shown, rather than a lot of additional faint stars beyond your aperture's light grasp. Negative printouts, showing black stars on white paper, are easiest to read at the eyepiece. A series of finder charts can be prepared: a wide-angle view comparable to the view through a finder telescope; a medium-magnification view of an area several degrees wide; and a detailed location chart showing the target area corresponding to the field of view through the eyepiece of the search instrument, and oriented to match the view through that eyepiece.

Using an 80 mm rich-field refractor with a low-power eyepiece or a pair of 60–80 mm binoculars, it is possible to locate more than 25 asteroids which brighten above 10th magnitude at opposition. For this sort of observing, binoculars are generally more satisfying to use than telescopes, since they deliver a "right way up" view and are comfortable to use, providing they are mounted well and the observer's neck is not craned too much when attempting to view objects high in the sky. Orientation is turned upside down through an astronomical telescope – north is at the bottom, and east is to the left. If a star diagonal is used, north is at the top but east remains at the left. It is essential that finder charts are printed with the correct orientation, and held so that the view matches that through the eyepiece.

On the finder chart, the asteroid's position should be circled faintly. At the eyepiece, the asteroid ought to be visible within this circle; mark the position of your suspected asteroid on the chart. For various reasons, however, your suspected asteroid may not be precisely where you think it ought to be. The most common reason is a discrepancy between the time of the calculated position of the asteroid and the time at which it is observed. Despite their great distance in the main asteroid belt beyond Mars, asteroids will appear to move among the background stars over a period of several hours. At around opposition, Ceres may appear to move against the background stars at the rate of around 30 arcseconds per hour – around 1/60 the field of view given by a standard Plössl eyepiece at a magnification of ×100 (to calculate an eyepiece's true field of view, simply divide the eyepiece's apparent field of view by its magnification). This

◀ *The motion of main-belt minor planets can be detected within a short space of time, as in this observation of Ceres made by the author at 00:30 and 01:15 UT with a 125 mm MCT at ×75, with a field of view of 40 arcminutes. Ceres moved 20 arcseconds.*

may not seem a great amount, but over a four-hour period it adds up to two arcminutes – around 1/15 the apparent diameter of the Moon. Another seemingly obvious but often overlooked factor in the apparent difference between an asteroid's observed and predicted position is the way the finder chart is held during an observation – so make sure its orientation precisely matches the view through the eyepiece.

If your finder chart does not show all the objects within the field of view, especially if there are several stars near the predicted position of your target asteroid, it will be necessary to draw in their positions as accurately as possible. Viewed through a telescope at a magnification of around ×50 – best use a wide-field eyepiece – the asteroid ought to change position noticeably with respect to the field stars after several hours, providing the asteroid is not near its stationary points at either side of the retrograde phase of its path. Of course, the asteroid will not be seen to move in "real time," but observations separated by several hours will be sufficient for it to have to betrayed its position. The asteroid will remain within the same general field of view over the space of a couple of days, so it can be checked on the next clear night.

Binoculars have such a wide true field of view that the asteroid's movement may not be obvious after a few hours, unless the asteroid is particularly near to a background star or forms a readily identifiable pattern with nearby stars. If it lies directly in between two other stars, or makes an equilateral triangle with them, for example, then any movement of the asteroid will be far easier to detect.

Advanced studies

It is an interesting exercise to attempt to discern the disks of the four largest asteroids. Ceres, the largest, measures a respectable 1.4 arcseconds in apparent diameter at opposition – making it more than half the apparent diameter of Neptune, and large enough to be resolved as a disk through a good 150 mm telescope under steady seeing conditions. It is

TABLE 14 OPPOSITIONS OF CERES 2006–2015				
Date	Constellation	RA	Dec	Magnitude
2006 Aug 12	Piscis Austrinus	21h 48m 20.3	−27° 52′ 05″	7.2
2007 Nov 10	Cetus	03h 07m 03.1	+08° 04′ 15″	6.9
2009 Feb 25	Leo	11h 02m 06.0	+24° 20′ 23″	6.4
2010 Jun 19	Sagittarius	17h 48m 29.1	−25° 27′ 23″	7.0
2011 Sep 17	Cetus	00h 01m 10.1	−17° 14′ 27″	7.2
2012 Dec 18	Taurus	05h 45m 33.3	+25° 14′ 39″	6.6
2014 Apr 16	Virgo	13h 52m 19.3	+03° 28′ 58″	6.5
2015 Jul 25	Microscopium	20h 27m 48.7	−30° 11′ 09″	7.1

TABLE 15 OPPOSITIONS OF VESTA 2006–2015				
Date	Constellation	RA	Dec	Magnitude
2006 Jan 06	Gemini	07h 06m 52.4	+22° 51′ 11″	6.2
2007 May 31	Ophiuchus	16h 32m 26.2	−14° 03′ 50″	5.2
2008 Oct 30	Cetus	02h 32m 20.4	+03° 34′ 51″	6.1
2010 Feb 18	Leo	10h 18m 42.8	+19° 46′ 10″	5.8
2011 Aug 06	Capricornus	21h 06m 59.1	−23° 07′ 17″	5.4
2012 Dec 09	Taurus	05h 07m 54.2	+17° 42′ 20″	6.2
2014 Apr 13	Virgo	13h 45m 23.1	+02° 47′ 50″	5.4
2015 Sep 29	Cetus	00h 39m 18.1	−08° 50′ 21″	5.9

TABLE 16 OPPOSITIONS OF PALLAS 2006–2015				
Date	Constellation	RA	Dec	Magnitude
2006 Jul 04	Hercules	18h 27m 13.9	+23° 18′ 26″	8.8
2007 Sep 04	Pegasus	22h 25m 56.9	+04° 00′ 46″	8.4
2008 Nov 30	Columba	05h 17m 50.0	−31° 56′ 42″	7.0
2010 May 01	Serpens Caput	15h 37m 43.3	+23° 53′ 10″	7.8
2011 Aug 01	Sagitta	19h 55m 07.9	+17° 19′ 08″	8.8
2012 Sep 24	Cetus	00h 24m 20.5	−07° 33′ 26″	8.0
2014 Feb 27	Hydra	09h 44m 15.7	−09° 57′ 14″	6.3
2015 Jun 12	Hercules	17h 32m 31.5	+25° 32′ 50″	8.5

TABLE 17 OPPOSITIONS OF JUNO 2006–2015				
Date	Constellation	RA	Dec	Magnitude
2007 Apr 10	Virgo	13h 27m 14.6	+01° 06′ 20″	9.5
2008 Jun 13	Ophiuchus	17h 29m 05.6	−04° 32′ 10″	9.7
2009 Sep 21	Pisces	23h 59m 44.2	−03° 50′ 35″	7.4
2011 Mar 12	Leo	11h 29m 11.2	+03° 50′ 29″	8.8
2012 May 20	Serpens Caput	16h 02m 46.0	−03° 02′ 13″	9.8
2013 Aug 05	Aquarius	20h 41m 19.2	−04° 56′ 22″	8.6
2015 Jan 31	Hydra	08h 31m 14.0	+03° 47′ 26″	7.8

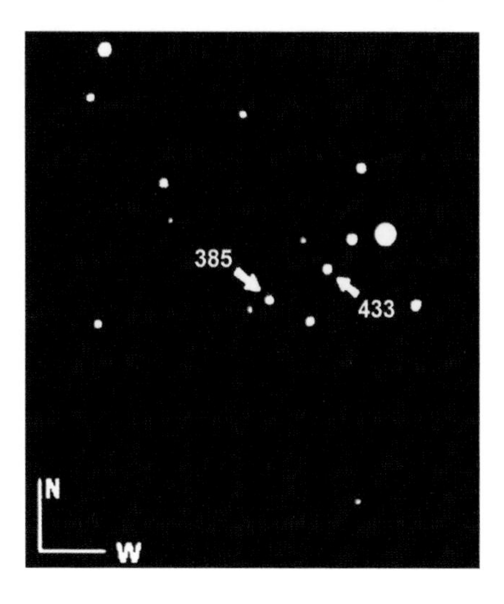

◀ *Minor planets (433) Eros and (385) Ilmatar, both faint 11th-magnitude objects, separated by about 2 arcminutes. CCD image by John Fletcher using a 250 mm SCT. January 25, 2003.*

not entirely beyond the bounds of possibility for a very keen-sighted visual observer to be able to discern vague surface features on the disk of Ceres through a 150 mm telescope under excellent seeing conditions, using the highest practical magnification. Using telescopes of 150 mm or larger, few visual observers experience any difficulty in discerning the disks of the Galilean moons – the largest of these, Ganymede, is just 1.6 arcseconds in apparent diameter at its largest, only slightly larger than Ceres. Since many observers have discerned detail upon all four Galileans, claims to have observed some kind of shading on Ceres ought to be seriously considered as valid observations.

At opposition, both Vesta and Pallas approach one arcsecond in apparent diameter – about the same apparent size as the smallest Galilean moon, Europa. Given good conditions and good optics, their disks can be comfortably resolved through a 200 mm telescope. Juno is just 0.6 arcseconds in apparent diameter at opposition, and is resolvable as a disk through telescopes larger than 350 mm.

Hunting down fainter asteroids requires larger instruments, dark skies and accurate finder charts. Given such conditions, for practical purposes, a 100 mm telescope can be used to hunt asteroids as faint as 10th magnitude; a 150 mm telescope will reveal 11th-magnitude asteroids, and a 200 mm telescope will clearly show asteroids of the 12th magnitude. Accurate predicted positions for asteroids, including fast-moving near-Earth objects, can be found online at the Minor Planet Center website (http://cfa-www.harvard.edu/iau/Ephemerides/index.html), where the user can compute a precise location for any asteroid by inputting the observer's terrestrial latitude, longitude and altitude, along with the time (in UT).

7 · JUPITER

Giant planet

Measuring 142,984 km across at the equator, Jupiter is by far the largest planet in the Solar System. It has an impressively fast rate of spin, revolving once on its axis in less than ten hours. Jupiter's rotational velocity at the equator is 45,260 km/h; the enormous centrifugal forces cause considerable bulging at its equator. As a consequence, the planet has a highly oblate form, with its polar axis measuring some 9267 km less than its diameter at the equator. Jupiter's oblateness is easy to discern at the eyepiece of even a small telescope. The giant planet is a vast ball of gas composed of 81% hydrogen, 17% helium and a few other elements such as methane, water vapor, ammonia and hydrogen sulfide, plus traces of more exotic gases; it is these gases which give a range of colors to the planet's cloud belts and other atmospheric features.

Dynamic atmosphere

Those Jovian features that are visible through the telescope eyepiece reside in the uppermost layers of the Jovian clouds, and undergo constant change. Jupiter's upper atmosphere does not appear to rotate as one uniform mass; it exhibits a differential rotation, with the equatorial regions rotating fastest and the polar regions moving slowest. Rotation periods on Jupiter range between around 9 hours 50 minutes in the equatorial region to around 9 hours 56 minutes elsewhere, with each zone and belt north and south of the equatorial zone having its own current, differing by fractions of a minute from its neighbors. Atmospheric turbulence springs up where two areas of different circulations shear past one another, giving rise to eddies. If the conditions are right, eddies develop into festoons, spots, ovals and other rotating systems, which can last for months, years, decades – even centuries, in the case of the Great Red Spot.

JUPITER: DATA	
Globe	
Diameter (equatorial)	142,984 km
Diameter (polar)	133,717 km
Density	1.33 g/cm^3
Mass (Earth = 1)	317.8
Volume (Earth = 1)	1320
Sidereal period of axial rotation (equatorial)	9h 50m 30s
Escape velocity	59.5 km/s
Albedo	0.73
Inclination of equator to orbit	3° 07′
Temperature at cloud-tops	125 K
Surface gravity (Earth = 1)	2.69
Orbit	
Semimajor axis	5.203 AU = 778.3 × 10^6 km
Eccentricity	0.048
Inclination to ecliptic	1° 18′
Sidereal period of revolution	11.86y
Mean orbital velocity	13.06 km/s
Satellites	over 60

Jupiter's orbit

Jupiter orbits the Sun every 11 years 315 days, ranging between a distance of 740,510 km at perihelion to 816,356 km at aphelion (an average distance of 778,330,000 km). Oppositions take place every 57 weeks, when the planet can shine as brightly as magnitude −2.9, if near perihelion, or magnitude −2, if near aphelion. Jupiter's orbital plane is inclined by just 1.3° to the ecliptic, and the planet appears to move slowly among the constellations of the zodiac at the rate of around 30° eastward per year. However, each of these constellations occupies a varying amount of space along the ecliptic, so Jupiter does not reach opposition in successive zodiacal constellations; it can skip a constellation, or reach opposition in a broad constellation such as Leo or Virgo in two successive years. Like all superior planets, Jupiter describes a looping motion against the background stars as the Earth overtakes the planet and our shifting line of sight causes it to appear to move in a retrograde motion for a period. This retrograde motion begins around 20 weeks after conjunction, when the planet appears to slow in its eastward path and become stationary. It then proceeds to move westward in a retrograde fashion for eight weeks prior to opposition and a further eight weeks following opposition, before slowing, becoming stationary again and finally continuing eastward, about 10° west of where it started from. Since Jupiter's orbit has such a low inclination, the actual retrograde loop that the planet traces is not very deep – sometimes the planet appears to go back over its former path, or describes a narrow zigzag path among the stars. Jupiter's motion can be discerned over a period of

◀ *Incredible detail within Jupiter's colorful, turbulent atmosphere can be seen in this image taken by the Cassini spaceprobe in December 2000, as it sneaked a view of the giant planet en route to Saturn. The Great Red Spot lies at upper left, while the small black shadow of the moon Europa lies at right on the same latitude.*

▲ *Jupiter, with three of the Galilean moons – Ganymede, Europa and Io (left to right), in an image by Dave Tyler. April 14, 2004, 21:58 UT.*

several hours if it is located in a field containing some background stars, which can be used to estimate its position. At around opposition, Jupiter appears to move westward at the rate of around 20 arcseconds per hour, amounting to one Jupiter diameter in around two and a half hours.

The Jovian system

Visible through binoculars, the four largest moons of Jupiter – Io, Europa, Ganymede and Callisto (called the Galilean moons after their discoverer, Galileo Galilei) – are all very large. Both Ganymede and Callisto are bigger than Mercury, with diameters of 5268 km and 4806 km respectively. Io is 3630 km across, making it slightly larger than our Moon. Europa is the smallest Galilean, with a diameter of 3120 km. Each orbits Jupiter very close to the planet's equatorial plane.

Fiery Io

Io is the innermost Galilean moon, orbiting Jupiter at an average distance of 421,600 km in a period of 1.77 days. It is a sizable rocky world, composed of silicates and sulfur.

Io's distinctive patchwork surface of orange, red, brown and black areas is caused by volcanism; it is the only planetary satellite whose surface is currently known to be volcanically active. Io's volcanoes shoot huge plumes of material up to 300 km above the satellite's

▲ *Mottled, multicolored Io, one of the Solar System's few volcanically active worlds.*

surface. This relentless volcanic activity is powered by a mechanism called tidal pumping, as Io's mantle is pulled around and frictionally heated by the tides raised by the combined gravitational pulls of Jupiter, Europa and Ganymede.

Cueball Europa

Although Europa is an icy world, like its Galilean siblings Ganymede and Callisto, its appearance is strikingly different – indeed, it looks quite unlike any world in the Solar System yet imaged. Europa orbits Jupiter at a mean distance of 670,900 km in a period of 3.55 days (twice as long as the orbital period of Io). The smallest of the Galilean moons, Europa has a crust made mainly of water ice. Spaceprobes have revealed a smooth white surface – one of the most topographically flat surfaces in the entire Solar System – striated with thousands of dusky brown lines. Fewer dark lines cross the satellite's polar regions, causing them to appear brighter than the rest of the surface; this polar brightening has been noted by numerous telescopic observers, particularly when the satellite has been observed in transit across a darker Jovian belt.

▲ *Europa's smooth icy surface is crossed by a multitude of dark narrow streaks.*

Europa's dark lines, some of which extend more than a thousand kilometers over its surface, represent parts of the crust which split apart, were filled from beneath by muddy slush, and then froze over once again. Its icy surface is relatively fresh, and few impact craters can be found – the largest, a bright ray crater called Pwyll, is about 50 km in diameter. It is probable that an ocean of liquid water exists between the outer ice crust and the satellite's rocky core.

Giant Ganymede

It is fitting that Jupiter, the largest planet in the Solar System, is orbited by Ganymede, the largest satellite in the Solar System. Ganymede orbits Jupiter every 7.15 days at an average distance of 1,070,000 km. It is composed mainly of ice and carbonaceous compounds. Its surface, covered with broad dark and light areas, bears witness to a complex history of crustal faulting, folding and melting, along with asteroidal impact. Broad light-colored bands cross Ganymede's surface for thousands of kilometers. On closer scrutiny these bands are composed of bright and dark ridges and valleys, evidence of crustal compression, deformation and faulting.

▲ *The surface of icy Callisto bears witness to thousands of sizable impacts.*

Cratered Callisto

Callisto, the outermost Galilean, orbits Jupiter at a distance of some 1,883,000 km in a period of 16.69 days. It is a world with an icy crust and a large core of rock and ice. Like Europa, Callisto probably has a subcrustal ocean of salty water. Its surface is the most heavily cratered in the entire Solar System, a fact made even more noticeable by the brightness of many of the younger craters, their ejecta systems and concentric fault rings. Valhalla, a bright crater some 300 km in diameter and surrounded by dozens of narrow concentric ridges to distances of more than 1000 km, is the most spectacular of Callisto's impact features.

▲ *Ganymede, the Solar System's largest satellite, has a complex patchwork surface.*

All of Jupiter's other satellites – more than 60 are currently known – are comparatively small, ranging from the irregularly shaped Amalthea (measuring 262 × 134 km) to minuscule bodies just a few kilometers across. None are visible visually through the telescope eyepiece, owing to their extreme faintness and the overwhelming glare of Jupiter. Amalthea's orbit, and the orbits of several of the other more sizable small Jovian satellites, lies close to Jupiter's equatorial plane and is nearly circular, but many of the smaller ones occupy a wide range of inclinations and have more eccentric orbits. All these minor satellites are irregular chunks of rock and/or ice, and are thoroughly pock-marked by impact craters. It is unlikely that these bodies were formed in the vicinity of Jupiter, as were the Galileans; they are more likely to have been asteroids and cometary nuclei that were plucked from deep space by Jupiter's gravity.

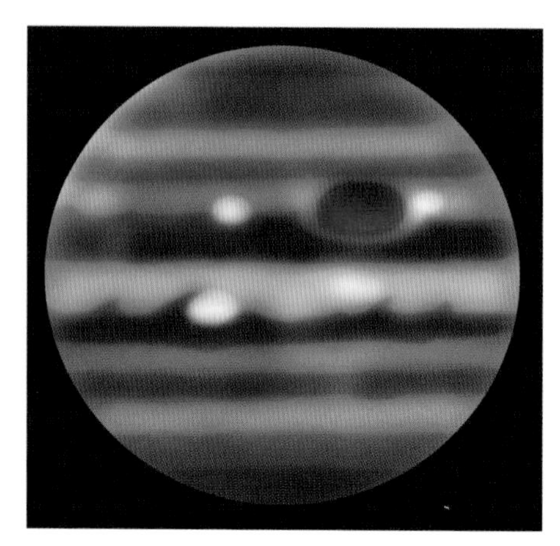

▲ *Graphic of Jupiter by the author, with the main belts and zones labeled.*

Observing Jupiter

Jupiter's giant status is apparent through the telescope eyepiece. Despite its incredible distance from the Earth – a maximum of more than 900 million km when observable in the early morning skies some time after conjunction, or in the evening skies prior to conjunction – Jupiter's disk always appears larger than 30 arcseconds in apparent diameter. At opposition the planet can be as close as 589 million km to the Earth, assuming a disk as large as 50.1 arcseconds across, which is easily large enough to be discerned as a small disk through a pair of steadily held 10 × 50 binoculars. Its four largest moons – Io, Europa, Ganymede and Callisto – are bright and easy to spot through binoculars (some eagle-eyed observers claim to be able to discern them with the unaided eye).

A telescope as small as 60 mm will reveal a few of Jupiter's darker cloud belts and some features within the clouds, such as prominent bright and dark spots and festoons looping off the North Equatorial Belt into the Equatorial Zone. Considerable atmospheric detail can be discerned through telescopes larger than 150 mm. Jupiter's ever-changing cloud features are fascinating to view through backyard telescopes.

Jupiter's equator is inclined just 3.1° to the plane of its orbit, so the planet always appears broadside-on, with its equator and atmospheric belts and zones running in nearly straight lines across a flattened disk. It is just possible to discern a slight curving of these features if one of the planet's poles is tilted at its maximum toward the Sun, but it is not obvious. Jovian phases are detectable too, even though Jupiter never appears less than 99% illuminated from the Earth. Its maximum phase can be seen around 14 weeks after conjunction, when the phase produces darkening on the planet's western side, and around 14 weeks following opposition, when the phase darkening is noticeable on Jupiter's eastern side. At opposition, of course, Jupiter appears 100% illuminated.

Jupiter does not have a solid surface, and its upper atmosphere is in a state of constant motion, so there can be no permanent features visible on the planet. Jupiter's rapid spin produces a prominent set of dark belts and light zones, parallel to its equator, and many of these features can easily be seen through a small telescope. The belts and zones appear to vary in color and intensity from year to year, but the most prominent of Jupiter's cloud features are usually the North and South Equatorial Belts. Features within the cloud belts and zones change in appearance on a virtually daily basis, as spots, ovals and festoons develop, drift in longitude, interact with one another and fade away. Features always remain within their own belt or zone, and they seldom drift much in latitude. A guide to Jovian activity appears later in this chapter.

Recording Jupiter

Disk drawings

Whole disk drawings of Jupiter depicting the relative broadness, positions and intensities of the planet's belts and zones, plus the finer detail discernable within them, make fine snapshots of the planet's appearance to the observer at any one moment in time.

Observing blanks for drawing the whole disk of Jupiter usually comprise an ellipse with a minor axis 1/15 smaller than its major axis – commonly a major axis of 63 mm and a minor axis of 59 mm. A cardboard template can be made to trace the ellipse onto plain paper; better still, a printed observing blank can be used. Drawings are best made with a medium-soft pencil.

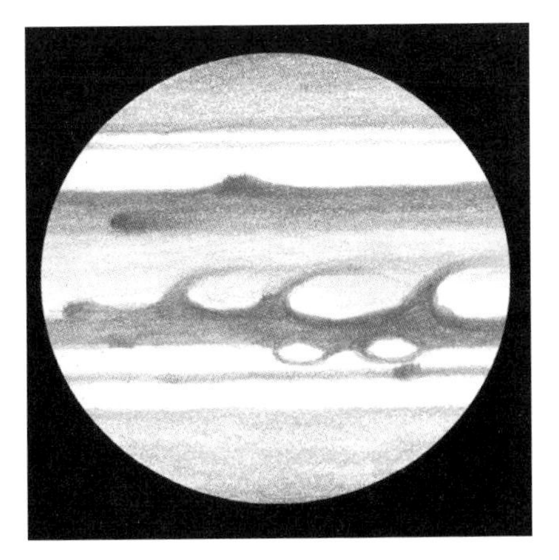

▲ *Disk drawing of Jupiter made by the author on July 25, 1999, at around 03:15 UT using a 225 mm Newtonian at ×250.*

Through a 100 mm telescope or larger, during periods of good seeing, an experienced observer may discern so much detail on the planet's broad disk that it may take a while to depict. Viewed at a power of ×150 – a sensible magnification to use with a 100 mm telescope for studying the planet on nights of average seeing – Jupiter's apparent diameter through the eyepiece is around 2° when near opposition. On the face of it, an area of 2° may not seem much, but bear in mind that this is around 16 times the area presented by the full Moon to the unaided eye, so there is a generous area around which to roam.

When Jupiter is first centered in the eyepiece, take in the scene and spend a while seeing which large or particularly prominent features are arrayed across the entire disk. Features on Jupiter appear to fade slightly toward the limb; some observers represent this fading with slight shading around the edge, while others choose to draw the belts, zones and any features near the limb slightly less distinctly than the rest.

After drawing in the readily discernable features, the observer is free to concentrate on one small, specific area at a time, attempting to tease out fine detail as seeing conditions allow. Detail requires some experience to see and fully appreciate, so an inexperienced observer may not see much more than the main belts and zones crossing Jupiter's disk.

Systems I and II

Jupiter's cloud system has a differential rate of rotation, the rotation period varying with latitude. Noting this, early observers correctly concluded that Jupiter's visible surface was not solid. Features within the planet's equatorial region take around 9 hours 50 minutes and 26 seconds to make a full rotation around the planet; features within this region, bounded by the north component of the South Equatorial Belt and the south component of the North Equatorial Belt, are referred to as System I features. Features located outside the equatorial region, including the Great Red Spot, rotate around Jupiter slightly more slowly, taking a period of around 9 hours 55 minutes and 29 seconds to make a circuit of the planet. Features in these regions belong to System II. The change in longitude of both systems over the same time period can be compared using Table 18.

Although there's a difference of just 5 minutes and 3 seconds per rotation between the two main Jovian atmospheric systems, the difference is compounded by each rotation. Features that were initially in line with each other on the planet's central meridian will have drifted about 30° apart in longitude after just 10 rotations of the planet. Within the space of just a month, the two features will be on directly opposite sides of the planet. This, however, is a very simplified way of viewing the Jovian atmosphere, but necessarily so. Jupiter's differential rotation is actually much more complex, and made more so by a number of jet-streams in the cloud belts which cause features to zip along at rates faster or slower than either of the two main atmospheric systems against which they are measured.

TABLE 18 LONGITUDE CHANGE OF SYSTEMS I AND II IN MEAN TIME INTERVALS

Time (minutes)	System I	System II	Time (hours)	System I	System II
1	0.6	0.6	1	36.6	36.3
2	1.2	1.2	2	73.2	72.5
3	1.8	1.8	3	109.7	108.8
4	2.4	2.4	4	146.3	145.1
5	3.0	3.0	5	182.9	181.3
6	3.7	3.6	6	219.5	217.6
7	4.3	4.2	7	256.1	253.8
8	4.9	4.8	8	292.7	290.1
9	5.5	5.4	9	329.2	326.4
10	6.1	6.0	10	365.8	362.6
20	12.2	12.1			
30	18.3	18.1			
40	24.4	24.2			
50	30.5	30.2			

Transit timings

Jupiter's rapid rotation gives observers the chance to make accurate timings of features as they cross the planet's central meridian (an imaginary straight line running down the center of the disk, joining both poles). By noting the time when a feature crosses the central meridian, the observer can discover the position of this feature in Jovian longitude, in either System I or System II coordinates, depending on which belt or zone it is in. Once the positions of Jovian features are known, they can be plotted on a map, and their movement in longitude relative to System I or System II coordinates can be traced over time.

System I and II coordinates are published in a variety of astronomical ephemerides; for example, the annual *BAA Handbook* gives the central meridian longitude of System I and System II coordinates in a daily ephemeris, for 00h UT. Some astronomical programs give the exact central meridian longitude for any time. Particularly noteworthy is the freeware program JUPOS, the aim of which is to collect precise positions of Jovian cloud features and to analyze them in drift charts in order to discover how their movements change in time. JUPOS is an amateur-astronomical project created by Hans-Jörg Mettig and Grischa Hahn; more information on JUPOS, including downloads for DOS and Windows PCs, can be found at http://jupos.org and http://jupos.privat.t-online.de/

Those who have never attempted to make transit timings may be under the impression that it would be practically impossible to achieve any sort of accuracy using the eye alone to judge when a feature is exactly on an unseen central line. However, regular transit timers achieve a remarkable degree of accuracy; despite using no scientific measuring devices (such as a micrometer), the experienced observer can consistently achieve a central meridian transit timing accuracy to within one minute – this amounts to just 0.6° of Jovian longitude.

Imagine that a large white oval has been spotted emerging onto the planet's eastern limb in the Equatorial Zone. It will take a couple of hours for Jupiter to rotate sufficiently for the preceding edge of this feature to reach the central meridian. Using an accurately set watch (preferably set to Universal time, UT, to prevent any errors when converting to and from local time), note the times when the preceding edge, center and following edge of the spot cross the central meridian. Jupiter has an equatorial diameter of 142,984 km and an equatorial circumference of 449,256 km. Assuming that the white oval is 10,000 km in diameter, it will cover almost 1/45 the planet's circumference, making it around 8° in Jovian longitude from preceding edge to following edge. Since the oval is in the Equatorial Zone, it belongs to System I, and will take 14 minutes to traverse the central meridian from one end to the other.

Strip drawings

One method of portraying Jovian features accurately over a period of several hours is to make a strip drawing of the planet. The planet's western limb (the limb over which features disappear) is drawn as a semi-ellipse, to the same scale as a regular disk drawing (with a major axis of 63 mm and a minor axis of 59 mm), and two horizontal parallel lines are drawn from the north and south points of the semi-ellipse. Inside this strip blank, the observer is free to depict features as they are brought onto the disk from the eastern limb and travel across the central meridian. On each side of the blank, transit timings of features can be indicated. This form of depicting Jupiter is much more accurate, more convenient and considerably less time-consuming than making a series of individual disk drawings to cover the same period of time. It helps to eliminate some of the ambiguities that occur when attempting to point out features on a regular disk drawing by means of note-taking alone.

Example observation of Jupiter – strip drawing plus transit timings

Date:	December 14/15, 1990
Time (UT):	23:04–04:00
Instrument:	100 mm refractor ×130
Observer:	Peter Grego

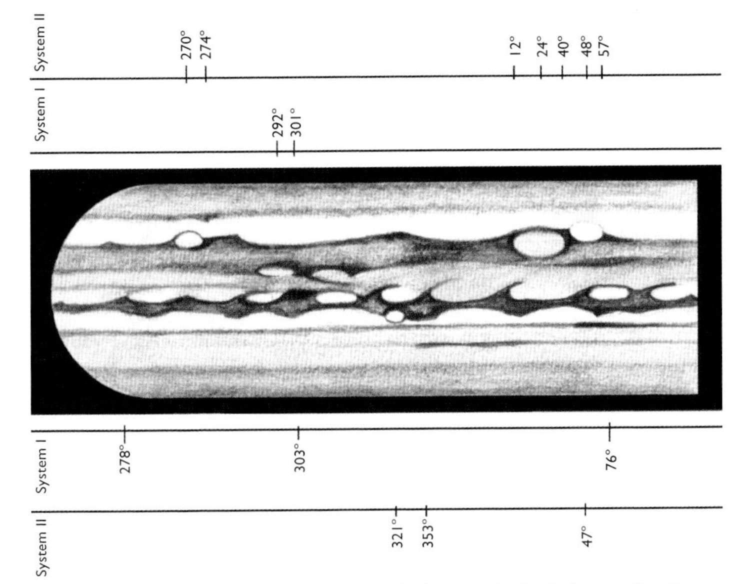

▼ *Strip drawing of Jupiter made by the author, showing half a rotation of the planet. It was drawn in a single session on December 14/15, 1990, between 23:00 and 04:00 UT using a 100 mm telescope. Some transit timed features are labeled. The Great Red Spot was prominent and easily visible.*

A total of 30 transit timings were made in a period of almost five hours (half a rotation of Jupiter). Just a few examples of timed features are given here and indicated on the strip drawing.

Location	Feature	Position	Brightness	CM timing	System I	System II
NEBs	Proj	C	d	23:10	278°	
STrZ	Oval	C	b!	23:13		270°
STrZ	Oval	F	b!	23:19		274°
SEBn	Oval	C	b!	23:32	292°	
SEBn	Oval	F	b!	23:47	301°	
NEBs	Proj	C	d!	23:50	303°	
NTrZ	Spot	C	b	00:37		321°
NNTB	Sect	P	d	01:12		353°
STrZ	RS	P	d	02:02		12°
STrZ	RS	C	d	02:21		24°
STrZ	RS	F	d	02:48		40°
NTB	Sect	P	d!	03:00		47°
STrZ	Oval	C	b!	03:01		48°
STrZ	Oval	F	b!	03:16		57°
NEBs	Oval	C	b!	03:28	76°	

TABLE 19 FEATURES	
Note that bright features can be noted by "b" (bright) or "b!" (brilliant), dark features followed by "d" (dark) or "d!" (very dark).	
Spot	Well-defined small spot, not too extended or oblong in shape.
Oval	Large well-defined oval-shaped area.
Bay	Large hollow which appears to "dent" the edge of a belt.
Nick	Small notch in the edge of a belt, frequently brighter than the adjoining zone.
Sect	Prominent section of a belt or zone.
Gap	Fairly extended, weakened or absent part of a belt.
Rift	Bright line crossing a broad belt, usually at an angle of 45° to 60°.
Area	Extended irregularly bordered bright area.
Strk	Pronounced oblong-shaped streak.
Shad	Shadow of a Galilean moon.
Moon	Disk of a Galilean moon.
Bar	Dark oblong object.
Fest	Dark filament or festoon that passes through a zone. One end of the filament or both ends of the festoon may emerge from dark belt condensations.
Proj	Dark projection on the edge of a belt. There are different shapes from rounded humps to tapering pyramidal objects.
Veil	Extended dark shading in a zone or in polar regions.
Dist	Dark integral or bridge-like structure which appears during SEB revivals.
Col	Column-like dark object in a zone, perpendicular or slightly inclined to it.

Jovian nomenclature and observing shorthand

So much is visible on Jupiter at any one time that it would be a grueling task to write down all that can be seen in standard longhand. When making notes and taking transit timings at the eyepiece, the system of abbreviations given in Tables 20 and 21 is suggested. Many of the feature types and their abbreviations have been based upon those outlined in the JUPOS program.

CCD imaging

CCD imaging has now become the most popular means of recording Jupiter – even more popular than making observational drawings at the eyepiece. CCD chips in ordinary webcams converted for astronomical use are sensitive enough to record Jupiter in detail and in color. Not only are CCD images an excellent means of capturing the appearance of the planet at a moment in time, they can also be used to derive accurate measurements of the latitude of Jovian belts and zones and the longitude of features.

Good CCD images of Jupiter, showing all of its major features, can be secured with equatorially driven telescopes as small as 100 mm. High-resolution imaging requires a telescope of at least 200 mm

TABLE 20 ABBREVIATIONS		
Belt/zone/feature	**Abbreviation**	**System**
North Polar Region	NPR	II
North North Temperate Zone	NNTZ	II
North North Temperate Belt	NNTB	II
North Temperate Zone	NTZ	II
North Temperate Belt	NTB	II
North Tropical Zone	NTrZ	II
North Equatorial Belt, northern component	NEBn	II
North Equatorial Belt, southern component	NEBs	I
Equatorial Zone, north	EZn	I
Equatorial Band	EB	I
Equatorial Zone, south	EZs	I
South Equatorial Belt, northern component	SEBn	I
South Equatorial Belt, southern component	SEBs	II
Great Red Spot Hollow	RSH	II
Great Red Spot	GRS	II
White Oval BA	WSBA	II
South Tropical Zone	STrZ	II
South Tropical Disturbance	STrD	II
South Temperate Belt	STB	II
South Temperate Zone	STZ	II
South South Temperate Belt	SSTB	II
South South Temperate Zone	SSTZ	II
South Polar Region	SPR	II

TABLE 21 POSITION	
Central Meridian	CM
Preceding edge	P
Center of feature	C
Following edge	F

▶ *Considerable atmospheric activity is shown in this CCD image of Jupiter, secured by Mike Brown on March 1, 2004, at 23:54 UT, using a 370 mm Newtonian. A small white spot is being swept along the northern margin of the Great Red Spot, giving the Red Spot Hollow a V-shaped appearance. A neat line of four white spots can be seen in the South Temperate Zone. The North North Temperate Belt appears to harbor great activity.*

aperture. Using an astronomical image-processing software package on the computer, short movie clips of Jupiter taken with a webcam can be chopped into individual frames; once these frames are automatically sorted by quality, the best of them are stacked and processed. The resulting image often shows far more detail than was immediately discernable visually at the eyepiece. Webcams can produce good results even when used in seeing conditions deemed too poor for visual observation.

To appear to be as representative of Jupiter as possible, the final image ought not to be over-processed. Limb darkening is far more evident on images than it is visually, and raising the limb brightness requires adjustment of brightness and contrast, plus unsharp masking. While image processing can be used to highlight certain vague and poorly defined features, obvious imaging artifacts are invariably produced. Artifacts produced by over-processing include bright rings around dark Jovian spots and Galilean shadows, and matrices of small crisscrossed lines. On close scrutiny of a CCD image, it is sometimes difficult to differentiate between real features and artifacts.

A guide to Jovian activity

Under good seeing conditions, a small telescope will show Jupiter's main belts and zones, along with a number of the smaller prominent atmospheric features. At least a 100 mm telescope is required to see Jupiter in any kind of detail. Most overt Jovian atmospheric activity takes place between the North Temperate Belt and the South Temperate Belt, and there is usually something immediately visible going on within or between the two main equatorial belts. Jupiter's dark belts and brighter zones vary in intensity from apparition to apparition; their intensity sometimes varies with longitude, with broad sections appearing darker or brighter than others. Occasionally a belt will fade into obscurity for an apparition or two, only to revive again. There is little obvious change in the latitude of the belts and zones, although their breadth varies from year to year – few observers take the trouble to make latitude measurements of the belts visually, as this can easily be ascertained by measuring photographs and CCD images of Jupiter.

In the following survey of Jupiter, the general appearance of each belt and zone is outlined, along with any recurrent phenomena that appear within them from time to time. Special mention is given to any interesting or unusual historic observations.

North Polar Region (NPR)

Extending southward to a latitude of around 50°N, the NPR usually appears dark and featureless through small telescopes. A 200 mm telescope will reveal occasional dark spots or streaks, with somewhat rarer anticyclonic white spots. The very northern part of the NPR often appears as a well-defined dark area of limited extent. The southern NPR usually directly adjoins the NNTB.

North North Temperate Zone and Belt (NNTZ and NNTB)

The NNTZ is an elusive feature which rarely stands out clearly from the north polar shading. Small anticyclonic white spots have occasionally been observed in the NNTZ during isolated apparitions. Sometimes the NNTB is clearly defined and darker than the NPR, but more often it appears irregular and difficult to trace. It tends to vary in intensity with longitude, and often takes on the appearance of a series of isolated darker sections whose preceding and following edges are poorly defined. On rare occasions the NNTB cannot be seen at all. Small, well-defined dark circular spots sometimes appear in the NNTB, moving in a rapid prograde fashion, along with small anticyclonic white ovals.

North Temperate Zone and Belt (NTZ and NTB)

Usually rather narrow but fairly conspicuous, the NTZ sometimes develops white spots and ovals which are large enough to occupy its whole width. After the NEB and SEB, the NTB is usually Jupiter's third most prominent belt; although it shows great variation in breadth and intensity over a period of time, it rarely exceeds the width of either equatorial belt. A great deal of interesting and unusual activity occurs in the NTB, much of which can often be observed through a 100 mm telescope. From time to time, over a period of a few months, the NTB becomes faint and indistinct in places, so that all that remains are a few dusky sections. NTB revivals tend to occur at a gradual pace, rather than in the frenetic fashion of SEB revivals (see below). Two clearly visible components, the NTBn and NTBs, usually make up the NTB. Features in the NTBn – usually dark elongated spots – tend to

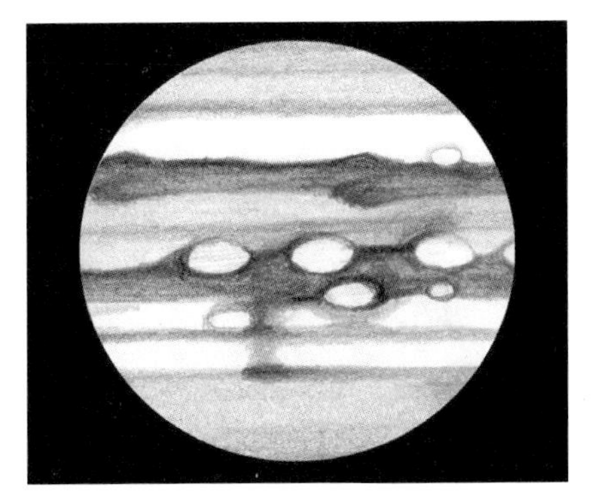

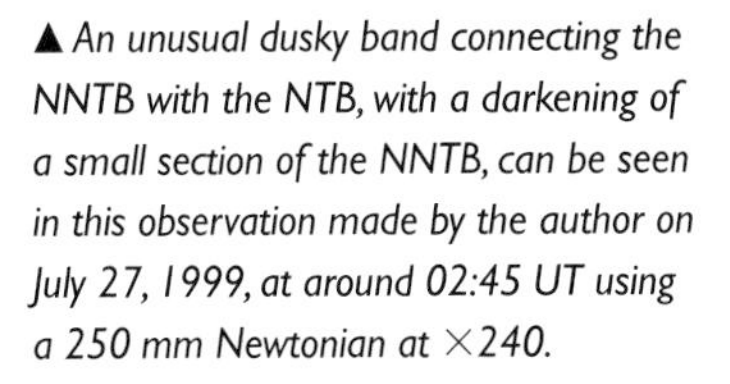

▲ *An unusual dusky band connecting the NNTB with the NTB, with a darkening of a small section of the NNTB, can be seen in this observation made by the author on July 27, 1999, at around 02:45 UT using a 250 mm Newtonian at ×240.*

move at the slowest rate of all Jovian features, in a period of around 9 hours 56 minutes. On occasion, dark projections from the NTBn have appeared to move directly in front of ovals visible in the NTZ, with little apparent interaction between the two features. In striking contrast with the NTBn, the fastest of all the Jovian atmosphere's jetstream currents rips through the NTBs. Some features observed here – mainly in the form of small dark features which project south into the NTrZ – have been carried along at rates faster than 9 hours 50 minutes, exceeding that of System I, and their motion relative to other more slow-moving features in the NTB is noticeable within the space of a few days.

North Tropical Zone (NTrZ)

Usually one of Jupiter's most conspicuous zones, the NTrZ displays little overt activity from year to year, although its brightness and breadth can vary considerably. The NTrZ may display a marked difference in brightness according to longitude, appearing brilliant on one hemisphere while being decidedly muted on the other. A narrow and rather faint dusky belt known as the North Tropical Band has sometimes been observed running down the center of the NTrZ.

North Equatorial Belt (NEB)

Usually easily visible through a 40 mm telescope, the NEB is Jupiter's darkest, most prominent and most consistently active belt. Over the decades it has been highly variable in width, ranging from a narrow line just a few degrees across, to a broad ruddy belt occupying 15° in latitude; it rarely fades from view altogether. Larger telescopes will reveal a wealth of fascinating activity within the NEB. It may appear

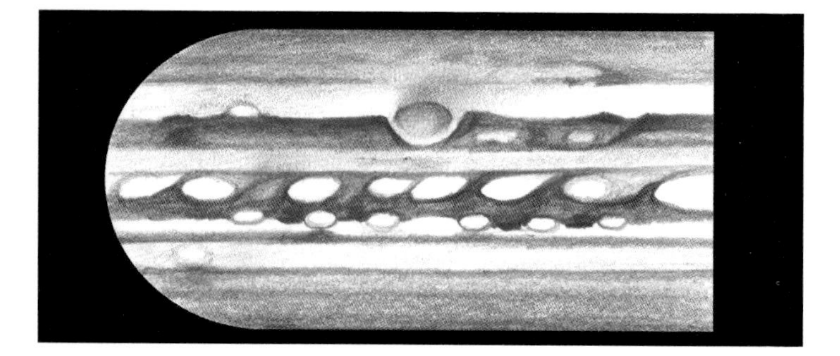

▲ Strip drawing of Jupiter made by the author between 23:00 and 03:00 on September 9/10, 1999, using a 250 mm Newtonian at ×240. Much activity can be seen, including a prominent GRS and dynamic activity on both sides of the NEB.

as a single belt of varying complexity, but it can sometimes display two distinct components (NEBn and NEBs) separated by a dusky zone; a narrow but distinct dusky third component has occasionally been noted along some sections of the NEB.

Small telescopes will usually reveal considerable scalloping and irregularity along the northern edge of the NEB; on closer scrutiny through larger instruments this breaks down into a series of dusky elongated spots and intensely dark streaks up to 15° in latitude, often quaintly referred to as "barges." These barges may be the darkest features visible on the entire Jovian disk, with the exception of satellite shadows, and a number of them are usually visible in each apparition. The dark barges sometimes occur in pairs separated by a bright spot. Bright circular anticyclonic spots and bays are frequently seen along the northern edge of the NEB; the spots may be distinctly brighter than the NTrZ. A long-lived storm called "White Spot Z" has been the largest and most prominent of the NEBn spots in recent years. When the NEBn is visible, all of these features are restricted to the northern edge of the belt, but when the NEB is broad and single they more often appear entirely nestled within the belt. Sometimes, features in the NEBn take on a most peculiar appearance, with the dark spots developing small bright centers; a line of such spots can give the distinct impression of the links on a chain.

Every few years, the NEB expands northward, giving the belt a much broader appearance, around 15° wide and extending to around 22°N or higher. These NEB expansion events usually begin with an accumulation of dark material on the edge of the NEBn which spreads in longitude to form a distinct dark boundary to the belt, an appearance which might last for an apparition or longer.

Bright white clouds sometimes develop deep within the NEB; in the space of a few days, the different wind currents between the NEBn and the EZn shear these clouds into prominent bright sections known as NEB rifts. NEB rifts can be straight or curved (one in 1928 was affectionately known as the "banana"), and can extend for tens of degrees. Typically they persist for several weeks.

Dark projections looping south and west from the NEBs into the EZ are known as festoons. They are nearly always present, clearly defined, and are frequently prominent enough to be seen through a small telescope. Viewed through telescopes of 200 mm or larger, festoons are noticeably blue in color. Some festoons are angled high into the EZ, and appear like dark shark's fins; others narrow and loop back toward the NEBs, some even rejoining it some distance to the east. Festoons frequently border bright plumes within the EZn; small telescopes often give them the appearance of large ovals.

Equatorial Zone (EZ)

There's always some sort of activity visible within the EZ, a generally bright area occupying Jupiter's equatorial regions. Much of this activity comprises dark festoons emanating from the NEBs, and bright plumes in the EZn stretching from the southern border of the NEBs. Other detail is often glimpsed within the EZs, with the occasional development of dark or bright circular spots. In 1999 a major storm known as the South Equatorial Disturbance erupted in the EZs; it soon spread around the globe along its latitude, and activity was visible into 2001.

When at its brightest, the EZ is white, but in some apparitions its tone is decidedly muted, and cursory observation through a small telescope on these occasions gives the impression that Jupiter has one single vast equatorial band. The EZ is not usually centered precisely above Jupiter's equator; it often occupies more of the northern equatorial latitudes than the southern latitudes. Its total width varies from around 10° at it narrowest to 18° at its widest. Occasionally a narrow Equatorial Band (EB) is visible in places, more or less coincident with Jupiter's actual equator, but it rarely forms a complete feature around the entire planet. During some apparitions the EB is dark and prominent, forming a band with which the festoons from the NEBs join.

South Equatorial Belt (SEB)

During some years, the SEB supplants the NEB as Jupiter's darkest and most prominent belt. It is certainly the most active belt on the planet and is prone to periodic disturbances, fadings and revivals, phenomena that are fascinating to follow through the eyepiece. The SEB is more often noticeably double than the NEB; its two components, the SEBn and SEBs, are usually more widely spaced than the two components of the NEB, and they are separated by a light-colored SEB Zone. The spacing between the SEBn and SEBs is not always visible at all longitudes, and in these locations the SEB appears as one single dark wide belt. Numerous dark spots and light rifts appear in the SEBn.

Among the most spectacular of events visible in Jupiter's atmosphere are fadings and revivals of the SEB. During a period of a few months the SEB can appear to diminish in intensity, leaving behind a faded SEBn. At the same time the Great Red Spot (GRS), an occupant of the STrZ bordering the SEB, darkens considerably. This peculiar appearance of Jupiter can last for a couple of apparitions, when, quite unexpectedly and within the space of a few days, a pair of spots forms – one intensely white and the other dark – in the SEBn, some 5° preceding the GRS. The bright spot is so prominent that it has been reported to appear to project from the planet's limb; this is a contrast effect between the bright spot and the limb fading, rather than a real indication of an enormous altitude. Within the space of several days, dark material spews out from the site of these spots, spreading rapidly eastward as it is carried along the SEBs jetstream and westward along the SEBn jetstream. A long dark filament of material spreads around the planet in the SEB's latitude, and a multitude of small dark condensations appears in both SEB components. Other spots erupt from the position at which the SEB disturbance originated, and within a few weeks, the SEB has revived, displaying numerous white spots, ovals and dark condensations. As the SEB regains its former intensity, the GRS usually fades.

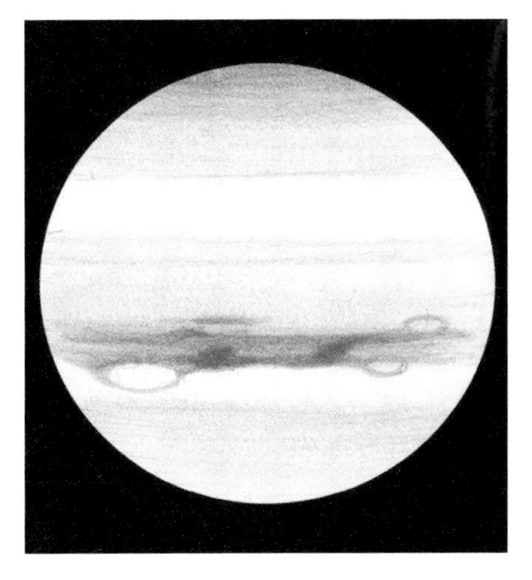

▲ *Jupiter's SEB has faded so much that it is barely visible in this observational drawing made by the author on August 28, 1989, using a 100 mm MCT at ×120.*

The portion of the SEB following the Great Red Spot usually displays a considerable degree of turbulence. Although this post-GRS disturbance appears much like a calm sea churned up in the wake of some great ship, this is not analogous to its actual cause. Instead, the turbulence takes place when bright spots (which originally formed in the SEB, around 50° following the GRS) encroach upon the following edge of the GRS and are ripped apart as they attempt to pass to its north along the Red Spot Hollow. A number of these spots are usually visible during a Jovian apparition. Outbreaks of white spots are also periodically observed around 100° in longitude following the position of the GRS. The spots propagate from a mid-SEB location and slowly advance toward the GRS, where they are quickly shredded apart.

The Great Red Spot (GRS) and Red Spot Hollow (RSH)

Systematic observations of Jupiter have been made for more than a century, but records of features observed on the giant planet date back to the mid-17th century. Perhaps the nearest thing that Jupiter has to a permanent feature is the famous Great Red Spot (GRS), a giant anticyclone in the planet's STrZ, so large that it could easily swallow the Earth. Having wandered in longitude around the planet since at least the mid-19th century, the GRS varies in intensity from year to year, ranging from a barely discernable gray smudge to a sharply defined brick-red oval which is easily visible through small telescopes.

In 1878 the GRS grabbed Victorian astronomers' attention because of its sheer size and intensity. Measuring some 34° in longitude – an elongated oval around 40,000 km wide and 13,000 km broad – its deep red color was discernable through a small telescope. The GRS displays variations in size from year to year, but it seems certain that, overall, it is slowly shrinking; its length is now some 30% smaller than it was back in the 1890s, during the golden age of visual Jovian studies. In 2005 the GRS measured some 18° in longitude. Its color has been described as ranging from "pale pink" to "brick red." During SEB revivals (see above) the GRS appears to darken as the SEB fades, and it fades away after the SEB revives.

Far from being a bland homogenous oval, the GRS can display considerable complexity when viewed through a 200 mm telescope under good seeing conditions. Its edge is almost always clearly defined, the preceding edge often being the darkest. Within the spot, several alternating light and dark rings can sometimes be discerned. A peculiar aspect of the GRS, noted in several apparitions, is an apparent elongation of both preceding and following edges, like the bows of a ship; this strange and puzzling appearance has never been satisfactorily explained.

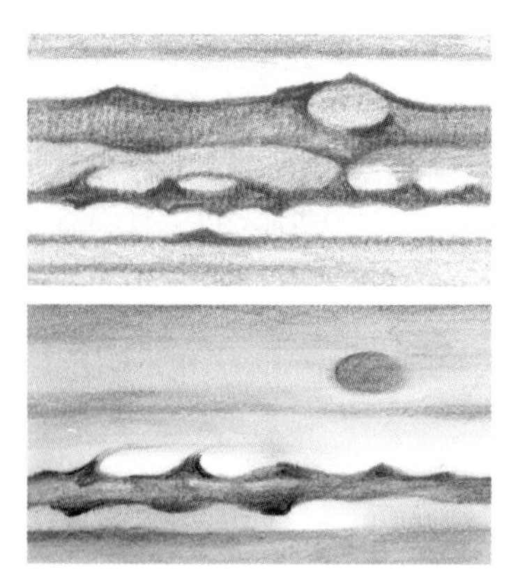

▲ *Jupiter's Great Red Spot varies somewhat in size, shape and intensity from year to year, as does activity within the planet's belts and zones. These two observational drawings compare the appearance of Jupiter on April 7, 1991 (top) and February 19, 1993 (bottom), both were made by the author with a 100 mm telescope.*

When at its faintest, the outline of the GRS is often impossible to trace with certainty, as its tone blends in with the surrounding STrZ. Sometimes it is an ill-defined gray haze, with the only indication of its presence being the Red Spot Hollow (RSH), a deep bay in the southern edge of the SEB. Faint wisps of dusky material sometimes extend from either edge of the RSH, curving across the STrZ to meet the STB, giving it the appearance of a well-defined giant light-colored oval, in which the GRS itself may not even be visible.

South Tropical Zone (STrZ)

Lying between the STB and SEBs, the STrZ is famous for being the domain of the GRS. In addition, this zone has given rise to a number of atmospheric upheavals known as South Tropical Disturbances. In 1901, the first observed (and greatest) South Tropical Disturbance disrupted the formerly clear STrZ. It started with a dark feature spanning the STrZ, which, after expanding in longitude, remained a permanent feature of Jupiter for more than 35 years. The preceding and following ends of the disturbance were concave, and a closed circulation developed within the region of the STrZ enclosed by the disturbance.

South Temperate Belt and Zone (STB and STZ)

During the 1930s, three prominent white ovals developed in the STB as a result of a disturbance in the STrZ. Designated BC, DE and FA, these ovals lasted for more than 60 years. After passing the GRS and drifting close to each other in 1997, ovals BC and DE collided and merged to form a single larger prominent oval, designated BE. After passing the GRS in late 1999, BE and FA collided in March 2000, creating a single prominent oval, BA, which currently remains a feature of the STB. Composed of a scattering of dusky streaks, the STB itself is not particularly prominent or well-defined, although it may revive at some point in the future.

South South Temperate Belt and Zone (SSTB and SSTZ)

Numerous anticyclonic storms are a regular feature of the SSTB; half a dozen or more are sometimes active at one time, and they can persist for a number of years. Though fairly small, they are bright and well-defined, and can be discerned with little difficulty through a 150 mm telescope under good seeing conditions. Small dark spots occasionally develop within the SSTB and the SSTZ.

South South South Temperate Belt and Zone (SSSTB and SSSTZ)

Larger instruments will pick up the brightest of the small white spots that occasionally develop within the SSSTB, an ill-defined area bordering the brighter SSSTZ to the south.

South Polar Region (SPR)

Usually a region that displays little obvious activity, the SPR occasionally hosts small bright or dark spots. The northern edge of the SPR is usually quite easy to trace, and subtle irregularities can sometimes be glimpsed along it.

▲ *Jupiter's flattened disk and its Galilean moons are bright enough to discern through steadily held binoculars. This image, taken by Cliff Meredith on February 15, 2002, at 22:08, shows (left to right) Io, Europa and Callisto.*

Galilean phenomena

Steadily held binoculars will reveal the four largest satellites of Jupiter – Io, Europa, Ganymede and Callisto – as four bright star-like points of light. Some people have claimed to be able to see one or more of these moons with the unaided eye; this is not an impossible claim, indeed all of them are bright enough to be visible to anyone with good eyesight under dark conditions if they were removed from Jupiter's glare.

A 150 mm telescope at high magnification will reveal the Galilean disks, although no really obvious surface detail can be made out. Io, Ganymede and Callisto are larger than our own Moon, and Europa is only slightly smaller.

Galilean transits, shadow transits, eclipses and occultations are frequent during every apparition, and they are fascinating to view. The Galileans also participate in a variety of mutual phenomena which are somewhat rarer, taking place every sixth apparition.

Disk and shadow transits

Sometimes the Galilean moons drift directly across the face of Jupiter. The satellites themselves, and more noticeably the black shadows they cast onto the planet, can be followed through a 100 mm telescope.

TABLE 22 GALILEAN SATELLITES

Galilean	Mean apparent diameter (arcsec)	Magnitude range
Io	1.1	5.3–5.8
Europa	1.0	5.7–6.4
Ganymede	1.6	4.9–5.3
Callisto	1.5	6.1–6.4

Each of the Galilean moons orbits very close to Jupiter's equatorial plane; since Jupiter's axial tilt is only 3.1° to the ecliptic, the moons appear to shuffle back and forth as they orbit Jupiter, transiting the planet when they move directly between the Earth and Jupiter, and then moving behind Jupiter on the other side of their orbit. Only Callisto is distant enough from Jupiter to avoid transiting the planet when its tilt is inclined sufficiently; this happens in three or four successive apparitions, separated by two or three successive apparitions in which Callisto transits occur.

Prior to opposition, the Galileans are preceded by the shadows they cast. At around western quadrature, Io's shadow is on Jupiter's central meridian when Io itself is just beginning its transit; Europa's shadow is about to exit as Europa enters the disk; and the shadows of both Ganymede and Callisto cannot be seen at the same time as the moons themselves are in transit. At eastern quadrature, the satellites precede their shadows. The apparent separation between each moon and its shadow diminishes as the apparition progresses. At opposition, the shadows appear to transit at around the same Jovian longitude as the disks themselves. Io, the nearest Galilean to Jupiter, is often superimposed upon its own shadow around the time of opposition, even when

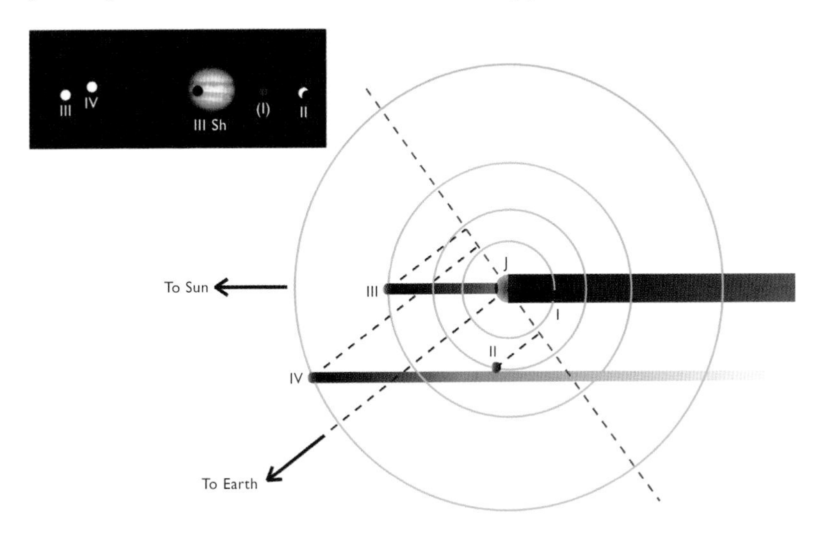

▲ *A hypothetical arrangement of Jupiter and the Galilean moons, showing a plan view from above the north pole of Jupiter and (inset) the view from the Earth. Satellite I (Io) is totally eclipsed by Jupiter's shadow; satellite II (Europa) is partially eclipsed by the shadow of satellite IV (Callisto), while the shadow cast by satellite III (Ganymede) is transiting the Jovian disk.*

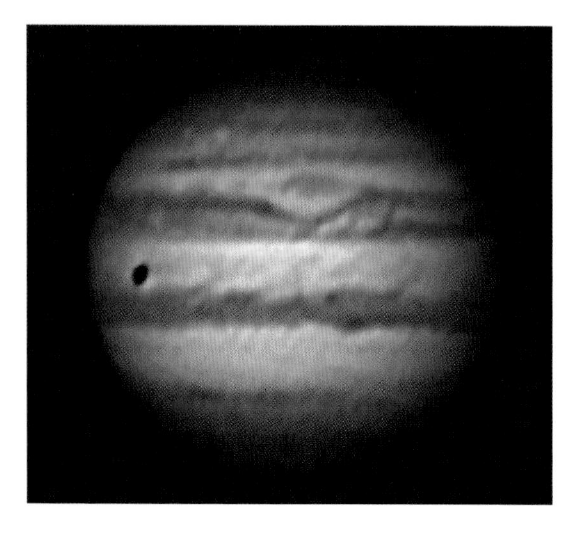

◀ *Callisto's shadow transits the Jovian disk on March 25, 2003, at 21:00 UT, imaged by Mike Brown using a 370 mm Newtonian and ToUcam Pro CCD camera.*

the planet is at its maximum tilt. The frequency with which the other moons are superimposed on their shadows decreases with their distance from the planet. Only in rare circumstances, when the plane of Jupiter's equator is aligned with the Earth, does the outermost moon, Callisto, appear against its own shadow on the Jovian disk. Following opposition, the shadows follow the disks of the moons casting them. Projected onto Jupiter's surface, the moons' dark umbral shadows (the umbra is the dark area completely cut off from direct sunlight) decrease in size with the satellites' distance from Jupiter. Io's shadow appears the largest and boldest, while Callisto's umbral shadow appears as a relatively tiny black dot surrounded by a very faint gray penumbra (the area where direct sunlight is only partly cut off by the moon's disk).

While in transit, the Galileans play a game of observational hide and seek, with their relative surface tones either blending in with, or standing out from, the Jovian atmospheric features against which they appear. Away from Jupiter, Io has a faintly orange hue, but once on the Jovian disk it assumes a pale gray color. Bright Europa is often lost against the brighter zones of Jupiter, while Ganymede and Callisto appear rather dark when set against a bright zone, but may disappear from view against a dark belt or feature of similar hue. All the moons stand out better around three months prior to or following opposition, near western or eastern quadrature, when Jupiter – and therefore the satellites themselves – display their maximum phase of around 99%. Phase darkening on either the preceding limb (before opposition) or following limb (after opposition) helps a great deal in defining the Galileans.

Under opposition circumstances, when the planet is fully illuminated and the planes of the Galileans' orbits are more or less aligned with the Earth, they all appear to transit Jupiter's equatorial region. When Jupiter's north pole is inclined toward us, the angled plane of the Galilean orbits causes the moons to transit against more southerly regions of Jupiter; conversely, when the planet's south pole is inclined toward us, the Galileans appear to transit more northerly Jovian zones and belts. Under such circumstances it is entirely possible for the outermost satellite, Callisto, to avoid a physical transit altogether yet still project a small shadow onto the Jovian disk.

From first contact on Jupiter's eastern limb to last contact on the western limb, fast-moving Io takes a maximum of around 2 hours 22 minutes to transit the Jovian disk; as it does so, it appears to travel around 90° in longitude against the Jovian background clouds. Europa takes a maximum of 2 hours 56 minutes to transit, and appears to move slightly more rapidly than the Jovian features beneath it. Ganymede transits in a maximum period of around 3 hours 46 minutes, and appears to move a little more slowly than Jupiter's rotation. Callisto transits Jupiter's equatorial region in a maximum period of around 4 hours 57 minutes; moving considerably more slowly than the planet appears to rotate, it tracks around 30° eastward against the planet's clouds during most of the transit (foreshortening effects are not taken into account). Of course, the shadows take about the same time as their satellites to transit the Jovian disk.

Eclipses and occultations

Jupiter casts an enormous shadow into the depths of space. Prior to opposition, a satellite approaching Jupiter on the far side of its orbit will be immersed into this shadow before reaching the planet's limb. These eclipses are not sudden events – over a period of several minutes, the eclipsed satellite slowly fades and vanishes from sight. Depending on the circumstances, it may emerge from the shadow before reaching Jupiter's western limb, or it may not be seen again until it reappears at the bright eastern limb of Jupiter. Following opposition, the Galileans undergo eclipses by Jupiter's shadow to the east of the planet.

Every sixth apparition, when the planes of the Galileans' orbits are aligned with the Earth, the moons can be observed to occult and eclipse one another. These mutual phenomena, many of which are quite rarely observed, are fascinating to watch. Often, there are a number of Galilean phenomena taking place simultaneously; lists of their occurrences are published in a variety of magazines and ephemerides, and they are also predictable using many good astronomical computer programs.

Oppositions of Jupiter, 2006 to 2015

When observing any astronomical object, its image begins to suffer noticeably from the effects of atmospheric murk and turbulence when it is less than about 20° above the horizon. Observers in northern temperate latitudes will find the years 2007 and 2008 frustrating for observing Jupiter, since the planet is located in the deep southern skies, only just edging above the southern horizon each night. During the same apparitions, observers in the southern hemisphere will be treated to a very high Jupiter, allowing a complete rotation of the planet to be observed in a single night around the time of opposition.

TABLE 23 OPPOSITIONS OF JUPITER 2006–2015

Date	Constellation	Declination	Equatorial diameter(arcsec)	Magnitude
2006 May 4	Libra	−14° 47′	44.7	−2.5
2007 Jun 6	Ophiuchus	−21° 54′	45.8	−2.6
2008 Jul 9	Sagittarius	−22° 29′	47.4	−2.7
2009 Aug 15	Capricornus	−15° 12′	48.9	−2.9
2010 Sep 21	Pisces	−02° 05′	49.9	−2.9
2011 Oct 29	Aries	11° 52′	49.6	−2.9
2012 Dec 3	Taurus	21° 20′	48.4	−2.8
2014 Jan 5	Gemini	22° 40′	46.8	−2.7
2015 Feb 6	Cancer	16° 27′	45.4	−2.6

8 · SATURN

The ringed planet

Saturn, its brighter satellites and its magnificent broad, bright rings present one of the most beautiful sights visible through the telescope eyepiece.

Measuring 120,536 km across at the equator, Saturn is the Solar System's second-largest planet. It is a gas giant made of about 93% hydrogen, 5% helium, small amounts of methane and water vapor, plus traces of a variety of other compounds, including ammonia.

Small telescopes reveal extensive dusky hoods over the planet's poles, plus two widely separated equatorial belts. Delicate banding within these belts is visible through larger telescopes under good seeing conditions. Lying parallel to Saturn's equator, the belts and zones are far less prominent and display less obvious activity than those of Jupiter. This is not surprising – Saturn orbits the Sun more than 650 million km further out than Jupiter, and receives considerably less heat to drive its atmospheric processes. Occasionally, subtle and short-lived localized features such as dusky knots and slight irregularities along the edges of the belts and zones can be discerned. Once every few decades a major atmospheric disturbance brews, producing a bright spot which dominates the equatorial regions for many months.

Observable activity in Saturn's cloud layers is muted by a high-altitude layer of haze formed by particles of ammonia ice. Beneath is a layer of ammonium hydrosulfide clouds, giving the planet its generally yellow hue, noticeable with the unaided eye.

All four of the Solar System's gas giants display bulging at the equator caused by centrifugal forces, but Saturn, with its polar diameter of 107,566 km being only 90% that of its equatorial diameter, displays the greatest degree of oblateness. Saturn's squat shape is far more striking when the planet's rings are presented nearly edgeways-on; when

SATURN: DATA	
Globe	
Diameter (equatorial)	120,536 km
Diameter (polar)	107,566 km
Density	0.69 g/cm^3
Mass (Earth = 1)	95.2
Volume (Earth = 1)	764
Sidereal period of axial rotation (equatorial)	10h 14m
Escape velocity	35.5 km/s
Albedo	0.70
Inclination of equator to orbit	25° 20′
Temperature at cloud-tops	95 K
Surface gravity (Earth = 1)	1.19
Orbit	
Semimajor axis	9.539 AU = 1427 × 10^6 km
Eccentricity	0.056
Inclination to ecliptic	2° 29′
Sidereal period of revolution	29.46y
Mean orbital velocity	9.65 km/s
Satellites	over 34

▲ *Saturn in a CCD image obtained by Dave Tyler on April 3, 2005, at 18:48 UT.*

the rings are tilted at an angle, they appear as elongated ellipses, making the planet's oblateness somewhat difficult to perceive. Saturn's equatorially distended form is produced by its rapid rate of spin – it revolves on its axis once every 10 hours 14 minutes. Although this is slightly slower than Jupiter's spin rate, Saturn is less dense and has a lower gravity. Indeed, the planet it is less dense than water, and if you had a large enough ocean, Saturn would float in it.

Saturn's orbit

Saturn orbits the Sun in a period of 29.46 years at an average distance of 1427 million km. Its orbit is the most eccentric of the four gas giants, ranging from 1348 million km at perihelion to 1506 million km at aphelion. Its orbital inclination of 2.5° to the plane of the ecliptic is the largest of the gas giants.

Oppositions occur around 13 days later each year. As Saturn slowly proceeds eastward its location at each opposition moves eastward at an annual rate of around 14°. Saturn undergoes retrograde motion for a period of a little more than two months either side of opposition, a total retrograde period of some 136 days.

Saturn's equator is inclined by some 25.33° to the plane of its orbit; this fact, combined with the slight inclination of its orbital plane to the ecliptic, means that the planet's maximum tilt toward us can exceed 27° on either side of its orbit, separated by just under 15 years. The degree to which Saturn is tilted is immediately noticeable because of the presentation of its rings.

Majestic rings

A good small refractor of 60 mm aperture will comfortably resolve Saturn and its rings clearly. Depending on the planet's phase and its tilt, the shadow of the rings on Saturn's globe and the globe's shadow on the rings can be discerned through a small telescope. Closer scrutiny will reveal that there are actually two main rings – an outer Ring A and an inner Ring B – separated by a dark narrow gap called the Cassini Division, after Jean Dominique Cassini who discovered it in 1676. A similar, though much narrower division lies near the outer edge of Ring A, called the Encke Division, after Johann Encke who observed it in 1837. In 1850 the presence of a broad but faint ring inside the B Ring was noted by William and George Bond; its translucent appearance was described by William Lassell as resembling "something like a crepe veil covering a part of the sky within the inner ring." Designated Ring C, observers still sometimes refer to this faint feature as the "Crepe Ring."

Numerous visual observers of the Victorian era reported that the main rings were made up of many fainter components, visible only under excellent seeing conditions through large telescopes. In view of the increasing complexity of the rings, along with the undeniable presence of the translucent Crepe Ring, the nature of the rings themselves began to be seriously questioned. In 1856 James Maxwell asserted that the rings could not possibly be solid; any such feature would be immediately torn up by Saturn's gravity, as the inner parts of the rings attempted to revolve around the planet faster than the outer parts. In fact, the rings lie entirely within a zone of gravitational disruption known as the Roche Limit – the critical distance from a planet at which a sizable solid body is gravitationally disrupted and torn apart by tidal forces. Maxwell realized that the rings must comprise "an indefinite number of unconnected particles." Proof of the particulate nature of Saturn's rings came in the form of spectroscopic observations made by James Keeler in 1895; his measurements of the Doppler shifts of various parts of the rings proved that their orbital periods increased with distance from Saturn. Each of the countless billions of constituent particles within the rings – from dust grain-sized motes to house-sized boulders – behaves as a tiny independent satellite, obeying Kepler's laws of planetary motion.

Anatomy of the rings

Ring A, the Encke and Cassini Divisions, Ring B and Ring C are the main ring components visible through amateur telescopes. From one side to the other, the main rings measure 273,600 km – that's one-fifth the diameter of the Sun. Starting closest to Saturn, Ring C is some 17,500 km wide, its inner edge located about 14,180 km above Saturn's cloud tops. Ring B is about 25,500 km wide, its outer edge lying 57,170 km above the cloud tops. After the Cassini Division, which forms a break 4700 km in width, Ring A extends to a distance of 76,470 km from Saturn's cloud tops. The center of the Encke Division is located 1100 km from the outer edge of Ring A.

Our knowledge of the ring system has increased beyond measure with the close-up views delivered by spaceprobes – notably the Voyager flybys of 1980–81 and Cassini, which commenced its orbital survey of the Saturnian system in 2004. They may look neat and relatively simple from afar, but the rings are actually made up of hundreds of individual ringlets of varying widths and densities. As observations have revealed increasingly more detail, a number of major rings and ring divisions have been added to the list, notably: Ring D – even fainter than Ring C inside which it lies; the narrow and somewhat enigmatic Ring F, just beyond the outside edge of Ring A; Ring G; and the extremely broad but exceedingly tenuous Ring E, which extends from the orbit of Mimas to a distance of almost half a million kilometers from Saturn.

It has long been known that the rings must be exceedingly thin, since they have been observed to disappear from view altogether on those infrequent occasions (every 15 years or so) when they have been

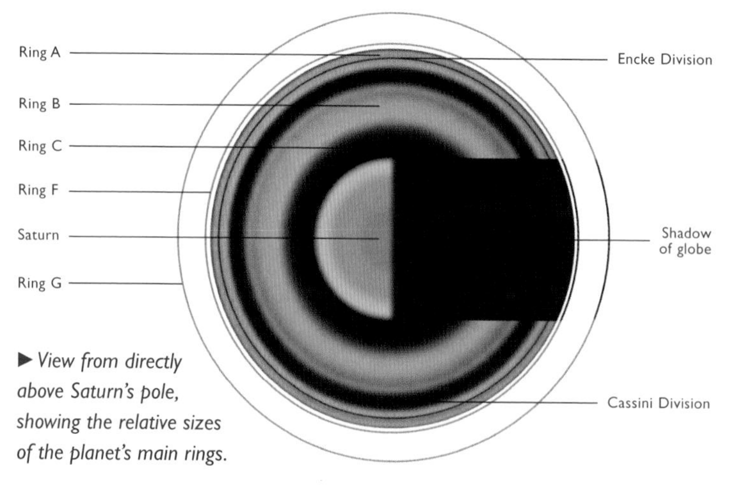

▶ *View from directly above Saturn's pole, showing the relative sizes of the planet's main rings.*

TABLE 24 THE SATURNIAN RING SYSTEM		
Feature	**Distance from Saturn's center (km)**	**Width (km)**
Cloud tops	60,330	
Inner edge of Ring D	66,900	7150 (Ring D)
Outer edge of Ring D	73,150	
Inner edge of Ring C	74,510	
Maxwell Division	87,500	270
Outer edge of Ring C	92,000	17,500 (Ring C)
Inner edge of Ring B	92,000	25,500 (Ring B)
Outer edge of Ring B	117,500	
Inner edge of Cassini Division	117,500	
Center of Cassini Division	119,000	
Outer edge of Cassini Division	122,200	
Huygens Gap	117,680	285–440
Inner edge of Ring A	122,200	
Center of Encke Minima	128,200	3500
Pan (satellite)	*133,583*	
Center of Encke Division	135,700	325 (Encke Division)
Center of Keeler Gap	136,530	35
Outer edge of Ring A	136,800	14,600 (Ring A)
Atlas (satellite)	*137,640*	
Prometheus (satellite)	*139,350*	
Center of Ring F	140,210	30–500 (Ring F)
Pandora (satellite)	*141,700*	
Epimetheus (satellite)	*151,422*	
Janus (satellite)	*151,472*	
Inner edge of Ring G	164,000	
Center of Ring G	168,000	8000 (Ring G)
Outer edge of Ring G	172,000	
Inner edge of Ring E	180,000	
Mimas (satellite)	*185,520*	
Brightest part of Ring E	230,000	
Enceladus (satellite)	*238,020*	
Tethys/Telesto/Calypso (satellites)	*294,660*	
Dione/Helene (satellites)	*377,400*	
Outer edge of Ring E	480,000	300,000 (Ring E)
Rhea (satellite)	*527,040*	

presented exactly edgewise on. At their thickest, the rings are less than one kilometer thick; to the same scale, if this paper represented the rings' thickness, the paper rings would be around the same diameter as a football stadium.

Radar observations have revealed that the particles within Saturn's rings range in size from tiny grains of dust, through pebble-sized lumps to house-sized chunks. The thickness of each of Saturn's ring components varies, and so too does the average size of the particles of which they are composed. Ring A is composed of a wide range of particle sizes, from small grains to objects up to 10 meters in diameter. Ring B, the thickest and widest ring, is not as homogenous as Ring A and displays hundreds of coarser ringlets; it is made of particles which range in size from 10 cm to around 1 meter. Spectroscopic observations have shown that the main ring particles are made of water ice, with a small proportion of rocky material. Rings A, B and F are slightly redder in hue than the other rings, suggesting that they contain a greater proportion of organic material (molecules made up of the most common elements found in living systems, that is, carbon, oxygen, nitrogen and hydrogen).

Although it is possible that Saturn may have had a ring system since its formation, the current configuration of the rings is not permanent. Under the gravitational influence of Saturn and its satellites, many of which actually lie within the ring system itself, the rings are undergoing continual, though gradual changes. Complex gravitational resonances between the so-called "shepherd" satellites of Atlas, Prometheus and Pandora and the ring particles themselves maintain the clean-cut edge of Ring A and the narrow Ring F just outside Ring A. In addition, these resonances produce ring regions of higher density as well as various divisions and gaps containing a paucity of material. Mimas, around 400 km in diameter and orbiting some 48,720 km from the outer edge of Ring A, is responsible for keeping the Cassini Division relatively clear of ring particles. The Encke Division is continually being swept clear of material by 19-km-diameter Pan, which orbits inside the Encke Division itself.

Today's telescopic view of the ring system is virtually indistinguishable from that perceived by astronomers of the 17th century, but over periods of hundreds of millions of years the ring particles are worn away by mutual collisions and eventually drift into Saturn's atmosphere. From time to time the rings are thought to be replenished by the gravitational disruption of larger satellites, or by passing comets which wander too close to Saturn. The current ring system is probably hundreds of millions of years old, and is likely to have been formed when a Saturnian satellite approached the planet within the Roche limit and was torn apart by tidal forces. If all the ring material were assembled together, it would form a large dirty snowball around 200 km in diameter – about the same size as the satellite Phoebe.

Saturn's satellites

Saturn is attended by a fascinating retinue of 34 known satellites, no fewer than seven of which are visible through a 150 mm telescope. Appropriately named Titan, by far the largest satellite of Saturn (and the second-largest satellite in the Solar System), is visible through binoculars. It has a diameter of 5150 km, and is the only satellite to possess a substantial atmosphere – mainly nitrogen, with some methane and traces of other compounds. Little of its surface can be seen because of its thick yellow cloud blanket, but its atmosphere displays some structure, including equatorial belts and polar collars; these features have apparently been glimpsed telescopically. A great deal of Titan's surface has been revealed by the Cassini spaceprobe, and images were secured by the Huygens lander in 2005. Spectacular images from Cassini have shown that Titan has weather and geology much like the Earth. Complex networks of drainage channels, which run from brighter highlands to darker, flatter regions, have been clearly imaged; the channels merge into rivers which drain into extensive lakebeds. There are only a few obvious signs of impact; Titan is such a dynamic world that fresh impact craters are quickly eroded.

Mimas, Enceladus, Tethys, Dione, Rhea and Iapetus are large spherical satellites composed mainly of water ice, and all are visible through a 200 mm telescope. Iapetus, Saturn's third-largest satellite, is the most

▲ *Saturn, with its brighter satellites Titan, Dione, Rhea, Enceladus and Tethys, observed by the author. March 16, 2003, at 23:20 UT, using a 150 mm refractor.*

TABLE 25 SATELLITES OF SATURN

(**Bold** indicates those satellites visible through a 150 mm telescope)

Satellite	Distance (km)	Average diameter (km)	Average magnitude
Pan	134,000	19	19
Atlas	138,000	28	18
Prometheus	139,000	94	15.8
Pandora	142,000	92	16.5
Epimetheus	151,000	114	15.7
Janus	151,000	178	14.5
Mimas	**186,000**	**392**	**12.9**
Enceladus	**238,000**	**520**	**11.7**
Tethys	**295,000**	**1060**	**10.2**
Telesto	295,000	30	18.5
Calypso	295,000	26	18.7
Dione	377,000	1120	10.4
Helene	377,000	32	18.4
Rhea	**527,000**	**1528**	**9.7**
Titan	**1,222,000**	**5150**	**8.3**
Hyperion	1,481,000	286	14.2
Iapetus	**3,561,000**	**1460**	**10.2–11.9**
Phoebe	12,952,000	220	16.4

observationally intriguing of them all. It varies in apparent brightness by 1.7 magnitudes at either elongation from Saturn, being brighter when west of the planet. This phenomenon was first noted by Iapetus' discoverer, Cassini, in 1671, who correctly suggested that it was caused by one half of Iapetus being more reflective than the other. Iapetus' leading side has a very low albedo, reflecting only 5% of sunlight, while the other hemisphere reflects up to 50% of sunlight. The less-reflective hemisphere, dominated by a region known as Cassini Regio, is covered by dark material likely to have been extruded onto Iapetus' surface through cryovolcanic eruptions.

Observing Saturn

Viewed with the unaided eye, Saturn shines with a steady, cream-colored hue. Its opposition brightness varies mainly with the angle at which the rings are presented to us, and to a lesser degree with the distance of the planet. At its brightest, when the rings are wide open and the planet is near perihelion, Saturn shines at magnitude −0.3, appearing brighter than every star apart from Sirius and Canopus. When the rings are presented edge-on, Saturn only manages to reach around magnitude 0.9, slightly brighter than Antares, the sky's 15th-brightest star.

TABLE 26 APPARENT DIAMETER OF SATURN'S RINGS AT OPPOSITION	
Feature	Mean apparent opposition diameter (arcsec)
Ring A (outer)	44
Ring A (inner)	38.7
Ring B (outer)	37.8
Ring B (inner)	29.2
Ring C (inner)	24.1
Saturn (equatorial diameter)	19.5

A pair of steadily held 10 × 50 binoculars will show Saturn as an elongated shape when the rings are open; the two brightest satellites, Rhea and Titan, are discernable through binoculars for most of their orbits when they are at a respectable angular distance from the planet.

From an observational viewpoint, Saturn's globe varies from a minimum apparent equatorial diameter of around 17 arcseconds, as it pulls out from conjunction and appears in the morning skies (or as it fades into the evening skies going toward conjunction near the end of an apparition), to a maximum of 21 arcseconds at a perihelic opposition, when its rings measure 46 arcseconds across.

A small refractor at a medium magnification will reveal the rings, the shadow of the rings on the globe and the globe's shadow on the rings (when the illumination conditions and tilt of Saturn permits). Detailed study of Saturn requires at least a 100 mm Maksutov–Cassegrain telescope or refractor, or a 150 mm reflector – these instruments will easily reveal the Cassini Division and the various belts on Saturn's globe. Really good views can be obtained through 200 mm telescopes and larger: seeing conditions permitting, the observer may be able to

▲ *Saturn, observed by the author on February 3, 1996, at 18:45, just nine days prior to ring plane passage. The rings were not visible through a 100 mm MCT. Satellites (from left to right) are Titan, Tethys, Rhea and Dione.*

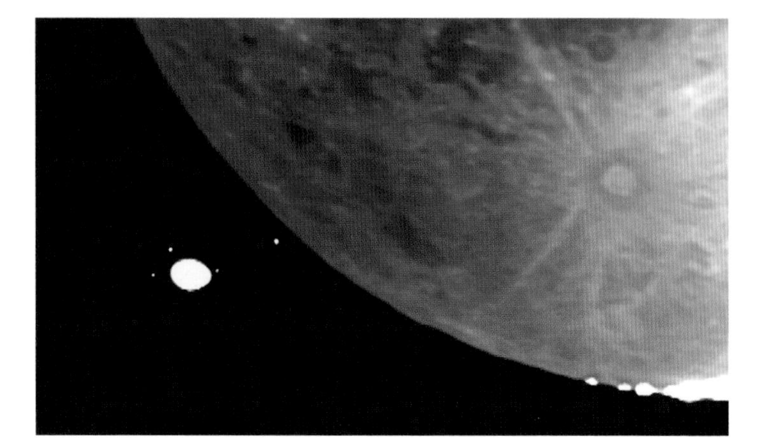

▲ *An overexposed image of the earthshine-lit lunar limb and Saturn; it was taken by John Fletcher a few minutes prior to the planet's occultation by the Moon on April 16, 2002. The image clearly shows Saturn's satellites Dione, Rhea, Tethys and Titan (from left to right).*

resolve the Encke Division, discern Ring C without difficulty, plus see no fewer than seven satellites, from 8th-magnitude Titan down to 12th-magnitude Mimas.

Spectacular spots

Saturn's globe appears fairly bland and serene for most of the time, displaying only occasional vague irregularities in the edges of its belts and subtle shading here and there. It has a generally yellow or cream-colored hue, though the equatorial belts may display an orange-brown tint, and the dusky polar regions can sometimes appear distinctly greenish.

Despite its apparent blandness, it is still worth paying attention to Saturn whenever it is visible – if only for 10 minutes out of each general observing session – since its atmosphere occasionally plays host to large white spots which suddenly well up and dominate the globe for several months. Any obvious detail on Saturn ought to be sketched and noted – and a central meridian transit timing (see page 146) of it made, if possible – since it may be a precursor to such a tumultuous atmospheric event as a great white spot. This startling phenomenon was observed in 1876, 1903, 1933, 1960 and most recently in 1990; it appears about every 30 years or so, when Saturn's northern hemisphere has been tilted strongly toward the Sun (note that Saturn's orbital period is 29.46 years).

An amateur astronomer, Stuart Wilber, discovered the 1990 white spot on September 24 of that year using a 250 mm reflector. In what was to become the most spectacular Saturnian atmospheric upheaval since 1933, the spot enlarged into a broad oval larger than Jupiter's Great Red Spot and was easily visible through small telescopes. After several weeks, 1700 km/h jetstreams in Saturn's upper atmosphere began to alter the spot's appearance: as it grew in longitude, it developed an extensive trailing area which eventually encircled the planet at its latitude. The newly operating Hubble Space Telescope was rescheduled to make observations of the storm; its high-resolution images showed considerable detail, notably scalloping and festoons, similar to those dark features that appear on the southern edge of Jupiter's North Equatorial Belt. By early November, the spot had spread into a broad, fairly homogenous bright belt which persisted into the following apparition.

Planetary scientists remain uncertain about the precise mechanisms responsible for Saturn's bright spots. They probably originate when the planet's atmosphere is heated and conditions are right for a large but well-defined atmospheric bubble to form deep beneath the cloud tops; when this bubble suddenly bursts into view, it forms a brilliant canopy of ammonia ice crystals which rise several hundred kilometers above the cloud tops. Although there is a generally acknowledged cycle of one major storm every 30 or so years, there is no reason to believe that a storm cannot take place at any time on Saturn – indeed, minor white spots and other atmospheric features are frequently captured on high-resolution CCD images by amateurs.

Rings

One of the most fascinating aspects of observing Saturn is the noticeably changing tilt of its ring system from one apparition to the next. Saturn's rings appear to be at their widest every 15 years or so – on one side of its orbit, the planet's north pole is tilted toward the Sun, and on the other side of its orbit its south pole is tilted toward the Sun. The maximum tilt amounts to some 27°, which is not quite enough for Saturn's globe to appear in perfect isolation from the rings, like a hole in a doughnut. At maximum tilt, Saturn's favored pole always appears to mask a small part of the far side of Ring B, while at the same time the near side of Ring B obscures a small segment of the opposite side of Saturn's globe.

When the rings are wide open, Rings A and B are easily visible through a small telescope, while the Cassini Division which separates the two rings may only be obvious as a dark crescent in the broadest part of each ansa, often being difficult to trace in its entirety through

▲ *The Hubble Space Telescope captured the changing tilt of Saturn between 1996 and 2000 (from lower left to upper right).*

small telescopes (the ansae are the parts of the rings visible on either side of the planet). Ring C is visible through a 150 mm telescope under good seeing conditions. A 200 mm telescope will comfortably reveal Saturn's major ring components, including the Encke Division and Encke Minima in Ring A, the Cassini Division, Ring B and Ring C.

Ring A appears a light gray color, and within it the Encke Minima – a slight darkening of the ring on the inside edge of the Encke Division – is more clearly visible than the Encke Division itself, which requires excellent seeing conditions to resolve as a thin black curve. Ring B is the most brilliant component of the rings; it displays a clearly defined duskiness around its inner third, a feature which inexperienced observers sometimes mistake for Ring C. Observers have sometimes noted a mottled appearance in the inner ansae of Ring B, and odd dusky "spokes" have been observed in both Rings A and B. Once considered illusory features, these spokes have been imaged by spaceprobes, vindicating the observations of astronomers like Barnard and Antoniadi. Contrary to what orbital dynamics might predict, the spokes appear to retain their radial integrity; their origin is thought to be associated with magnetic or electrostatic forces within the rings and their effect upon small susceptible ring particles.

Unlike Rings A and B, Ring C is faint and translucent, an effect caused by its being composed of widely separated ringlets. Where Ring C crosses Saturn, the globe behind it appears grayed-out – indeed, this effect has often been mistaken for a belt near the planet's equator. Steady seeing and good contrasty optics are essential for viewing the ansae of Ring C in the curved space between the globe and the inner edge of Ring B.

Following the maximum presentation of the rings, it takes around 7.3 years for Saturn's tilt to ease back so that its polar axis is perpendicular to our line of sight and the rings are presented edge-on. Since Saturn's ring plane is the same as that of its equator, on those occasions when the Earth and Saturn's equatorial plane become exactly aligned for a brief period, a phenomenon called a "ring plane passage" occurs. A single or triple ring plane passage takes place during any half-orbit of Saturn; in those apparitions when a single ring plane passage takes place, Saturn is located on the far side of its orbit and is so close to the Sun that it is difficult to observe. Apparitions featuring triple ring plane passages have one such event at around the date of opposition and one on either side of opposition, when the planet is around western and eastern quadrature. For a period of as little as a few hours on either side of a ring plane passage, the rings may vanish from sight altogether, even when viewed through quite large telescopes.

From our terrestrial vantage point, three factors influence the visibility of the rings when they are edge-on. First, when the Earth lies in the ring plane, the rings may be too narrow to be resolved through small instruments. Secondly, when the Sun is in the ring plane, neither face of the rings is illuminated; during such periods, the rings cannot be discerned visually even through very large telescopes. Third, when the Sun and the Earth are inclined very slightly to the ring plane but are on opposite sides of it, so that terrestrial observers are viewing the side of the rings that are not being directly illuminated by the Sun; visual observers have sometimes reported the faint visibility of the dark side of the rings on such occasions.

When Saturn's rings are presented edgewise, a number of subtle shadings may be observed in their extremities (at this time the section of the rings overlying the globe is exceedingly difficult to discern, though the shadow of the rings on the globe may be more easily seen). Some of these shadings may reflect the location of the main ring components; for example, the location of the outer edge of Ring B may appear as a brighter streak, while the location of the Cassini Division may be darker than the streak of Ring A. Distinct knots and condensations within the rings have been observed; they may not be symmetrical on either side of the globe, nor correspond with the location of the components of Rings A and B. The brightness of the rings on each side of the planet may be asymmetric; indeed, observers have reported the complete invisibility of the rings on one side of the planet while the rings were clearly visible as a narrow line on the other side.

The period of time in which Saturn's rings appear as a bright, narrow single line depends on the instrument used and the seeing conditions. Through a 200 mm telescope, the rings may appear as a single linear feature for several weeks, and may only be resolved as highly foreshortened curved lines around a month or so after the date of ring plane passage. Smaller telescopes have less ability to resolve the rings; a 100 mm telescope may not reveal the rings as anything but narrow featureless streaks for several months on end.

Saturnian shadowplay

A number of curious effects caused by the shadow of the globe on the rings and the shadow of the rings on the globe have been noted by visual observers. When the rings are edge-on to the Earth, and at the same time the Sun is inclined to the plane of the rings, the shadow of the rings appears to run across Saturn's equator, bisecting the planet; sometimes the rings cannot be discerned, giving Saturn a peculiar bivalved appearance. When the rings are edge-on to the Earth, the edge of the rings' shadow (if, indeed, the shadow is visible) nearest Saturn's equator invariably appears straight, but the shadow's far edge sometimes has a concave outline, appearing to broaden toward the planet's east and west limbs. The opposite effect has also been occasionally noted, however, with the shadow seeming to taper toward each limb.

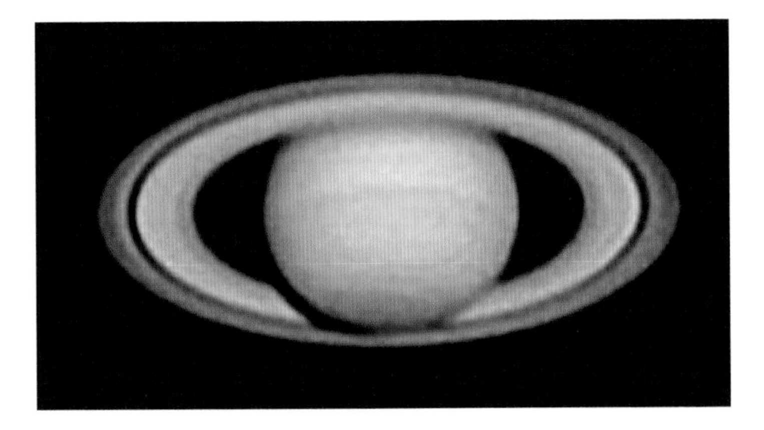

▲ Saturn, with its rings fully open. Note the dusky Ring C against the globe. This CCD image was obtained by Mike Brown using a 370 mm Newtonian.

As Saturn's rings open out, observers can follow the planet's changing appearance through each apparition as the shadow of the globe sweeps across the rings beyond. When Saturn becomes observable in the morning skies following conjunction, the shadow of the globe on the rings is noticeable adjoining the globe's western limb. Depending on the tilt of the rings, the globe's shadow can either completely cover a section of the rings, or it can appear as a curved shadow covering part of the rings. At opposition, the Sun, Earth and Saturn are in approximate line with one another; Saturn is fully illuminated, and casts its shadow directly away from our line of sight. Only when Saturn's rings are tilted sufficiently is the shadow of the globe on the rings discernable at opposition. After opposition, the shadow of the globe on the rings continues to proceed eastward. Measured from the planet's limb, the maximum breadth of the shadow cast onto the rings is greatest at western or eastern quadrature.

Where the shadow of the globe crosses the rings at the Cassini Division, observers have sometimes reported the curious effect of the shadow appearing to curve away from the planet at the junction between the outer edge of Ring B and the Cassini Division, giving the tapering point of Ring B a blunted appearance. Sometimes visible on photographs, the effect results from contrast between the dark shadow and the bright rings, and the inability to clearly resolve the tapering point of the illuminated portion of Ring B where it meets the Cassini Division. The effect is most noticeable on observations or images made with small or poorly focused telescopes and/or in less than perfect seeing conditions.

Other unusual shadow phenomena include: the rarely observed apparently convex profile of the globe's shadow on the rings; a pointed appearance to the top of the globe's shadow on the rings; the hazy outline of the edge of the globe's shadow; and a diffuse gray appearance of the entire shadow of the globe on the rings. Some of these effects have optical causes, while others are likely due to irradiation of light from Saturn and/or its bright rings into the shadow of the globe on the rings. A notable contrast effect, known as Terby's white spot, causes a small portion of the rings adjoining the shadow of the globe on the rings to appear markedly brighter than the rest; the effect is more pronounced through smaller telescopes.

Observing the satellites

No fewer than seven of Saturn's many satellites are visible through a 150 mm telescope. In order of distance from Saturn, they are: Mimas, Enceladus, Tethys, Dione, Rhea, Titan and Iapetus. Of these, Mimas is the most difficult to discern because of its faintness (mag 12.9) and its

◀ *Saturn and its brighter satellites imaged by Dave Tyler, December 7, 2004, 18:48 UT.*

proximity to the glare of Saturn; it zips around the ringed planet in a period of just 22 hours 37 minutes, and at maximum elongation it lies a mere 8 arcseconds from the outer edge of Ring A. Enceladus, shining at an average opposition magnitude of 11.7, is a little easier to view than Mimas. Its orbital period is 1 day 8 hours 53 minutes, and it is some 15 arcseconds from the edge of Ring A at greatest elongation. A 150 mm telescope will reveal both Mimas and Enceladus at elongation.

Tethys orbits Saturn in a period of 1 day 21 hours 18 minutes. Shining at magnitude 10.2, it can be seen without difficulty through a 100 mm telescope; at elongation, it lies around 25 arcseconds from the outer edge of Ring A. Dione is slightly fainter, with a mean opposition magnitude of 10.4, but its distance from Saturn at elongation – some 37 arcseconds – makes it just as easy to identify. Dione revolves around Saturn in 2 days 17 hours 41 minutes.

Rhea, the fifth satellite from Saturn in this list, orbits in a period of 4 days 12 hours 25 minutes, and at maximum elongation it lies some 1.3 arcminutes from Saturn's center. Rhea shines at magnitude 9.7 and can be easily seen through an 80 mm telescope at maximum elongation.

By far the easiest satellite of Saturn to identify, Titan shines at a mean opposition magnitude of 8.3 – bright enough to be discerned through binoculars. Its orbital period is 15 days 22 hours 41 minutes, and at maximum elongation it appears around 3 arcminutes from the center of Saturn. Titan's 5150-km-diameter disk appears some 0.7 arcseconds across and may be discerned in large telescopes using a high magnification, given good seeing conditions. From time to time, observations of vague detail within Titan's thick atmosphere have been made by experienced astronomers using large instruments. A notable series of observations was made by Bernard Lyot using the 600 mm refractor of the Pic-du-Midi observatory at magnifications of 1000 and 1250; they appeared to show an active, changing atmosphere, with an equatorial region that sometimes appeared bright or dark.

Iapetus, the furthest satellite in this list, orbits Saturn in a period of 79 days 7 hours 56 minutes. It is an interesting satellite which displays a marked difference in brightness on either side of its orbit. When west of Saturn, Iapetus is at its brightest, shining at magnitude 10.2; east of the planet, it is magnitude 11.9 (average opposition brightness). An unevenness in the satellite's surface brightness causes this phenomenon – one hemisphere is relatively bright, while the other is very dark.

Since the brightest satellites visible through amateur instruments orbit Saturn along its equatorial plane, the apparent paths that they take around the planet vary with its tilt. When the rings are edge-on, the satellites appear to shuttle back and forth across Saturn's disk and rings. Occasionally, two or more satellites can appear embedded along the rings like pearls on a necklace. Mutual phenomena between the satellites take place as they eclipse and occult one another. Also, the satellites and their shadows transit the planet's disk.

Saturn's tilt means that the various satellite phenomena cannot be observed during each apparition, as they can with Jupiter, but instead are restricted to around two apparitions on either side of Earth's passage through Saturn's ring plane. The number of observable mutual satellite phenomena, satellite and shadow transits increases nearer to the date of ring plane passage; at the other extreme, for around five years on either side of the planet's maximum tilt there are no observable satellite phenomena.

Only at times when the rings are edge on or the ring angle is very small are the satellites occulted or eclipsed. Titan appears as a dark body against the globe of Saturn when in transit, appearing almost as dark as its shadow, a phenomenon visible in small telescopes. Of all the satellites, only Titan and its shadow are large enough to be discerned on the globe of Saturn through a 200 mm telescope. When transiting Saturn's equatorial region, Titan (and its shadow) takes around six hours to move from the western to the eastern limb. From time to time, Saturn's other satellites move in front of the shadow of the rings on the globe, but Saturn's brightness drowns them out, rendering them unobservable.

Recording Saturn

Not only must Saturn's oblateness be depicted accurately when preparing an observing blank from scratch, but also its changing tilt, which alters the appearance of its rings. Including the globe, the two main rings and the Cassini Division, no fewer than five ellipses must be accurately drawn on an outline blank. Given the correct measurements, this can be accomplished fairly easily on a suitable graphics computer program. The dimensions are: Saturn's polar diameter –

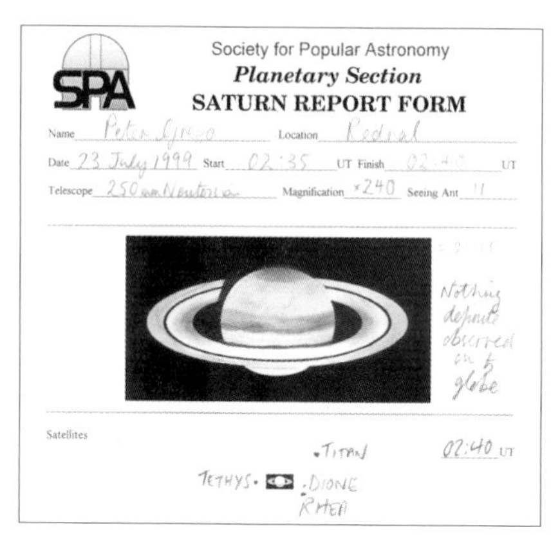

SPA

Society for Popular Astronomy

Planetary Section

SATURN REPORT FORM

Name Peter Grego Location

Date 23 July 1999 Start 02:35 UT Finish UT

Telescope 250 mm Newtonian Magnification ×240 Seeing Ant II

Nothing definite observed on the globe

Satellites

•TITAN 02:40 UT

TETHYS • :DIONE

RHEA

▲ *Observational drawing of Saturn made on a standard report blank. The observation was made by the author on July 23, 1999, using a 250 mm Newtonian.*

42 mm; Saturn's equatorial diameter – 44 mm; inner edge of Ring B – 67 mm; outer edge of Ring B – 87 mm; inner edge of Ring A – 90 mm; outer edge of Ring A – 102 mm. Instead of constructing the blanks themselves, most observers make use of outline blanks (made to approximately the correct tilt); these are supplied by the observing sections of a number of major astronomical societies, either as preprinted blanks, or print-it-yourself versions available from their Internet resources.

Having an accurate outline blank upon which to base an observation only helps so far; getting the correct curve of Saturn's belts and zones, the shadow of the globe on the rings and the rings on the globe, along with any fine detail within the rings themselves, is a matter for the observer to attend to. Most observers begin by sketching the more obvious shadows, darkening the Cassini Division, and, if appropriate, marking where the shadow of the rings on the globe can be seen through the Cassini Division. Next, the darker prominent features can be drawn – this may include toning down Ring A and the inner part of Ring B, the dusky polar hoods and Saturn's belts. Faint features, such as any delicate shading in the belts and zones, subtle detail within the rings, along with Ring C and even Ring D, should be left until last, once the main features have been depicted.

Transit timings

As with Jupiter (see page 111), the longitudes of features in Saturn's atmosphere can be calculated by making timings of their passage across the planet's imaginary central meridian. With the eye alone, an accomplished observer can make transit timings with an accuracy of one minute, equating to just 0.6° of Saturnian longitude. Tables giving the longitude of Saturn's central meridian at 0h UT on a daily basis are published in such sources as the *BAA Handbook*. Like Jupiter, Saturn has two main systems of atmospheric rotation. System I includes the whole equatorial zone, between the south component of the North Equatorial Belt and the north component of the South Equatorial Belt.

System I has a period of 10 hours 14 minutes 13 seconds (a rotation rate of 844° per Earth day). System II includes the rest of the planet, and has a period of 10 hours 38 minutes 25 seconds (a rotation rate of 812° per Earth day).

Intensity estimations

By estimating the relative intensity of parts of the globe and rings of Saturn on a regular and systematic basis, the observer can build up a useful record of the planet's changing appearance. While most changes on Saturn may not have quite the dramatic visual impact of those on Jupiter, in the long term it can be seen that there is a great range in the actual intensity of Saturn's belts and zones. Saturn intensity estimates are made on a scale from 1 to 10, where 1 is the brightness of Ring B just inside the Cassini Division, and 10 is the intensity of the dark background sky. Regular observers may build up enough skill to be confident in applying fractions of ½ in their estimations.

Filter work

Saturn is amenable to observations with filters, but its relatively low brightness means that only telescopes larger than 200 mm are suitable for this kind of work. Filters recommended for Saturn observing are Wratten 25 (red) and Wratten 44a or 47 (blue). By comparing intensity estimates made in white (unfiltered), red and blue light, a difference in the relative intensities of the belts, zones and rings can be revealed, indicating color tints. Additionally, the widths of the belts and zones may appear to vary when viewed through different color filters. An unusual phenomenon visible through filters, known as the bi-color aspect of the rings, reveals a discrepancy in the intensity of opposite ansae when viewed in red and blue light; the cause is unknown, but it is a real effect which has been verified on CCD images.

TABLE 27 OPPOSITIONS OF SATURN 2006–2015

Date	Equatorial diameter (arcsec)	Tilt	Magnitude
2006 January 27	20.5	−18.9°	−0.2
2007 February 10	20.3	−13.9°	0.0
2008 February 24	20.1	−8.4°	0.2
2009 March 8	19.8	−2.6°	0.5
2010 March 22	19.6	3.2°	0.5
2011 April 4	19.3	8.6°	0.4
2012 April 15	19.1	13.7°	0.2
2013 April 28	18.9	18.2°	0.1
2014 May 10	18.7	21.9°	0.1
2015 May 23	18.5	24.5°	0.0

9 · URANUS

Uranus, the seventh major planet from the Sun, is a gas giant measuring 51,118 km across at the equator, around four times the Earth's diameter. Its atmosphere is composed mainly of hydrogen (82%) and helium (14%) gas, with a relatively large proportion of methane (around 2%) which has frozen out into icy clouds. The planet's striking light turquoise hue is attributable to these clouds, since methane absorbs red and yellow light. Beneath Uranus' atmosphere, some planetary scientists think that a mantle of ice perhaps 6000 km thick surrounds a rocky core with a diameter of around 10,000 km.

URANUS: DATA	
Globe	
Diameter (equatorial)	51,118 km
Diameter (polar)	49,947 km
Density	1.29 g/cm^3
Mass (Earth = 1)	14.53
Volume (Earth = 1)	62.18
Sidereal period of axial rotation (equatorial)	17h 14m (retrograde)
Escape velocity	21.3 km/s
Albedo	0.51
Inclination of equator to orbit	97° 52′
Temperature at cloud-tops	55 K
Surface gravity (Earth = 1)	0.79
Orbit	
Semimajor axis	19.22 AU = 2871 × 10^6 km
Eccentricity	0.046
Inclination to ecliptic	0° 46′
Sidereal period of revolution	84.01y
Mean orbital velocity	6.81 km/s
Satellites	over 15

Herschel's planet

Uranus was discovered in 1781 by William Herschel during a routine sweep of the sky using his 155-mm self-made reflector. Then in the constellation of Taurus, Uranus was found to be just visible with the unaided eye, appearing as a star of magnitude 5.6, but its tiny disk could clearly be discerned. Uranus' discovery almost doubled the known size of the Solar System.

Rotation and orbit

At perigee, Uranus approaches the Sun to 2.73 billion km; at its furthest, it reaches a distance of more than 3 billion km – more than 20 times the distance of the Earth from the Sun. Uranus takes a little more than 84 years to make one circuit around the Sun in an orbit inclined by just 0.77° to the ecliptic.

Uranus rotates in a period of 17 hours 14 minutes in the opposite direction to most other planets. Its axial tilt is 97.86°, so that the planet effectively orbits the Sun "lying on its side." This produces interesting seasonal effects. When Uranus' north pole points toward the Sun, its northern hemisphere basks in continual sunlight; the main satellites appear to revolve around the planet in near-circular, clockwise

▶ *Uranus, observed by the author through a 150 mm refractor ×188 on July 28, 2002.*

orbits. Twenty-one years later (a quarter of the planet's orbit around the Sun) the planet's polar axis has assumed a position at right angles to the Sun; all parts of the planet revolve into sunlight during the 17-hour-long day, and the satellites appear to move back and forth across the disk of Uranus in the plane of its equator. After another 21 years, Uranus' south pole has rotated to point sunward, and the northern hemisphere is in continual darkness; the satellites appear to revolve in near-circular orbits in an anticlockwise direction. In 2006 the south pole is proceeding toward the planet's edge, and in 2007 Uranus will be sideways-on. Afterward, the planet's north pole will gradually turn toward the Sun, until in 2030 we will view it from almost directly above the pole.

▼ *Subtle shading at Uranus' Earth-pointed polar region, observed by the author on July 23, 1984, at 20:45 UT using the 400 mm telescope of the University of Birmingham Observatory.*

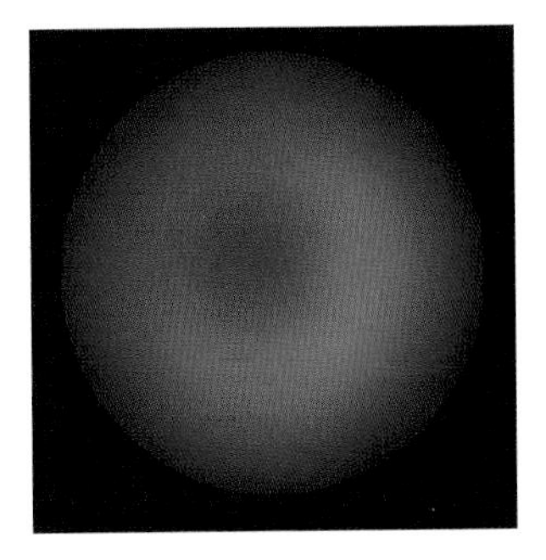

Rings

Uranus, like the Solar System's three other gas giants, has a ring system. It was discovered as recently as 1977 when a team of astronomers observing a stellar occultation by Uranus, noticed the star wink on and off five times prior to the main occultation, and five times following it, as the unseen rings briefly hid the star from view. By comparing their results, the observers concluded that Uranus'

◀ *Uranus, its rings and some of its satellites, imaged by the Hubble Space Telescope.*

rings are exceedingly dark and very thin – a picture since confirmed by Voyager 2 and large telescopes. In all, there are nine major rings interspersed among broader dusty tracts encircling the planet. Although the rings are far too faint to observe visually, it is possible that a well-equipped CCD imager could capture a trace of the rings. Future stellar occultations by Uranus and its rings will, from time to time, be observable through amateur instruments.

Satellites

Of the 15 known satellites of Uranus, the four biggest and brightest were discovered visually at the telescope eyepiece. Titania and Oberon were found in 1787 by William Herschel, just six years after he had discovered Uranus itself. Ariel and Umbriel, somewhat closer to Uranus and rather faint, were discovered in 1851 by William Lassell.

Observing Uranus

Oppositions take place around 4 days later each year, when it reaches magnitude 5.6 – bright enough to be seen with the unaided eye, given keen sight and dark skies. After 2010 the planet nudges north of the celestial equator, reaching its highest northern declination in 2034; it remains in the northern celestial hemisphere until 2053.

Uranus' mean opposition diameter of 3.8 arcseconds (3.96 arcseconds at maximum, when Uranus is near perihelion) is sufficiently large for its disk to be perceived through quite small telescopes at medium to high magnifications. Its green hue is noticeable through small telescopes, too. Uranus often presents a rather bland, featureless disk. Vague belts and zones can sometimes be glimpsed, but nothing as certain as those features visible on Jupiter or Saturn. Since Uranus' belts and zones are parallel to the planet's equator, they can appear to cross the planet in a variety of ways, depending on the angle at which the planet's polar axis is tilted. In 2007 the belts and zones will appear as linear stripes, since the polar axis is at right angles to our line of sight. In 2030, however, our view from almost directly above the planet's north pole may show a central dusky polar hood surrounded by the concentric rings of the belts and zones.

Observing the satellites

Four of Uranus' main satellites are within the well-equipped amateur's domain. Both the inner satellites of Uranus, Ariel and Umbriel, are difficult objects to observe because of their proximity to Uranus and their faintness. The mean opposition magnitude of Ariel is 13.5, while Umbriel is a whole magnitude fainter. William Lassell discovered the pair through a 600 mm reflector, but they can be glimpsed through slightly smaller telescopes. Ariel's orbital period is 2 days 12 hours 29 minutes; Umbriel's period is 4 days 3 hours 28 minutes. A 200 mm telescope will show Titania (magnitude 13.7) and Oberon (magnitude 13.8), which, respectively, take 8 days 16 hours 56 minutes and 13 days 11 hours 7 minutes to orbit Uranus.

► *A comparison between the apparent sizes of Uranus and Jupiter by Dave Tyler.*

Experienced CCD imagers using much smaller instruments (say a 150 mm telescope) will certainly have little difficulty in securing images of all four of Uranus' main satellites, and perhaps even 17th-magnitude Miranda.

Recording Uranus

Using steadily held binoculars, careful observational drawings of the starfield in which Uranus appears will enable the planet's slow movement to be discerned over a period of a few weeks and months. Little of scientific importance is expected to come from routine visual observations of Uranus, but surprises can frequently happen in astronomy. It is not entirely beyond the bounds of possibility that Uranus may develop a bright spot or some sort of major atmospheric feature prominent enough to be seen through amateur telescopes. A vigilant amateur is likely to capture such an event – either through visual observation or on CCD images – before the professionals.

Uranus' disk shows a lesser degree of oblateness than the disks of Jupiter and Saturn. It can be accurately represented on an observing blank using the information provided by any number of astronomical planetarium programs, but most observers are content to use a circular disk of 50 mm diameter when making observations of the planet itself. It is worth being aware of the planet's tilt and its orientation through the eyepiece, just to clarify the position of any features which may be observed; for example, if a small, distinct spot is discerned, it is useful to know whether it is approaching or receding from the planet's central meridian, in order to make a transit timing.

TABLE 28 OPPOSITIONS OF URANUS 2006–2015

Date	RA	Declination	Constellation
2006 Sep 05	22h 58m	−07° 31′	Aquarius
2007 Sep 09	23h 12m	−05° 59′	Aquarius
2008 Sep 13	23h 27m	−04° 26′	Aquarius
2009 Sep 17	23h 41m	−02° 51′	Pisces
2010 Sep 21	23h 56m	−01° 16′	Pisces
2011 Sep 26	00h 11m	+00° 19′	Pisces
2012 Sep 29	00h 25m	+01° 54′	Pisces
2013 Oct 03	00h 40m	+03° 29′	Pisces
2014 Oct 07	00h 55m	+05° 04′	Pisces
2015 Oct 12	01h 09m	+06° 37′	Pisces
2016 Oct 15	01h 24m	+08° 09′	Pisces

10 · NEPTUNE

Neptune is a gas giant measuring 49,528 km across at the equator, making it slightly smaller than Uranus. Its atmosphere, composed mainly of hydrogen (84%) and helium (12%) gas, with a relatively large proportion of methane (around 2%), is very similar to that of Uranus. Methane is an effective absorber of red and yellow light, and the planet appears distinctly blue.

Neptune's atmosphere is far more dynamic than that of Uranus. Prominent light and dark bands encircle the planet, and within them are to be found sizable anticyclonic spots, atmospheric plumes and festoons, which appear at a range of heights. Among the atmospheric phenomena imaged during Voyager 2's brief snapshot of Neptune in 1989 were several large dark spots – the largest of which, dubbed the "Great Dark Spot," measured 15,000 km across – and numerous bright high-altitude cirrus-like streaks. Although these features appear to be of a more transient nature than those of Jupiter, observations by the Hubble Space Telescope and large Earth-based telescopes show that there is always large-scale activity taking place in Neptune's atmosphere.

A calculated discovery

Unlike Uranus, Neptune was not discovered by accident. Careful measurements of Uranus' position suggested that its orbital path was not consistent with that predicted by Newton's laws of gravitation. Either the laws were wrong or Uranus was being gravitationally perturbed by a large unknown planet lying beyond it; most astronomers chose to believe the latter. In 1846 the gifted mathematician and astronomer Urbain Leverrier calculated the probable position of this mystery planet, and communicated his predictions to Johann Galle of the Berlin Observatory. Wasting no time in beginning his search, Galle discovered Neptune on September 23, 1846, using the observatory's 225 mm refractor; the planet was

NEPTUNE: DATA	
Globe	
Diameter (equatorial)	49,528 km
Diameter (polar)	48,686 km
Density	1.64 g/cm^3
Mass (Earth 5 1)	17.14
Volume (Earth 5 1)	57.67
Sidereal period of axial rotation	16h 07m
Escape velocity	23.5 km/s
Albedo	0.84
Inclination of equator to orbit	28° 18′
Temperature at cloud-tops	55 K
Surface gravity (Earth 5 1)	0.98
Orbit	
Semimajor axis	30.06 AU = 4497 × 10^6 km
Eccentricity	0.0097
Inclination to ecliptic	1° 46′
Sidereal period of revolution	164.79y
Mean orbital velocity	5.43 km/s
Satellites	13

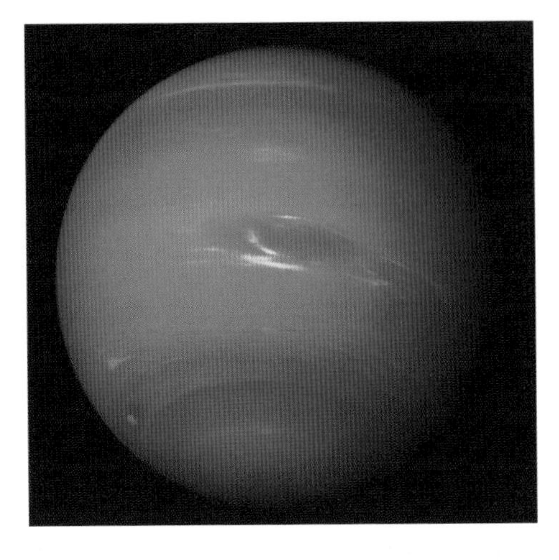

◀ *Neptune and its great dark spot, imaged by Voyager 2 in 1989.*

within just one degree of its position as predicted by Leverrier. Shining at magnitude 7.8 and measuring 2.3 arcseconds in apparent diameter, it was too small and faint to be seen without optical aid.

By looking at historical observations prior to Neptune's discovery, it was found that a number of astronomers had actually seen and recorded Neptune, but failing to discern its tiny disk, they had believed it to be a star. Indeed, Neptune was very close to being discovered at the very dawn of the telescopic era. Observing Jupiter on December 28, 1612, Galileo noted two faint stars in the same telescopic field of view; one of these happened to be Neptune, though it was far too tiny be resolved as a disk through his small refractor. On observing Jupiter again on January 27, 1613, Galileo noted that the two nearby stars appeared to be closer. He was right – Neptune had moved 15 arcminutes further west – but the astronomer believed that his first observation may have been a mistake, so he never followed it up. In fact, Jupiter occulted Neptune on the evening of January 3, 1612. If Galileo had been fortunate enough to have observed this event, he would have deduced that the "star" actually had a sizable apparent diameter, since Neptune took more than half an hour to disappear behind the Jovian limb. Stars, on the other hand, are occulted far more rapidly due to their minuscule apparent diameters.

Orbit and rotation

Neptune orbits the Sun in a near-circular path, inclined by 1.77° to the ecliptic, at an average distance of 4.5 billion km – 30 times the distance of the Earth from the Sun. At such a vast distance, Neptune takes 164.8 years to orbit the Sun – on May 29, 2011, the planet will have completed exactly one orbit since its discovery. Neptune's equatorial plane is inclined 28.3° to the plane of its orbit, and each Neptunian season lasts more than 41 years. The planet's south pole is currently around its maximum tilt toward the Sun, and throughout the first half of the 21st century the planet's tilt will gradually decrease, so that its equatorial plane will be aligned with the Sun by 2046. Afterward, the planet's north pole will gradually turn toward the Sun.

Neptune's rotation period of 16 hours 7 minutes is the slowest of the gas giants, and its small degree of oblateness produces a difference of around 900 km between the planet's equatorial and polar diameters. Neptune's oblateness is not immediately apparent through the eyepiece, especially when its axis is tilted at a high angle toward the observer.

Rings

Neptune's rings were discovered by occultation studies from the Earth in 1983. They comprise four exceedingly narrow, faint components which are thought to be made up of particles thrown out by meteorite impacts with Neptune's satellites. Displaying a degree of clumpiness, the rings are less uniform than the systems around the three other gas giants; indeed, the uneven distribution of material in the rings produces asymmetric occultation effects, leading some early researchers to conclude that the rings were made up of a series of disjointed arcs of material.

Satellites

Prior to Voyager 2's flyby in August 1989, Neptune was known to have only two satellites – Triton and Nereid. Voyager and later Earth-based images have revealed 11 more Neptunian satellites, and there are probably many more small ones awaiting discovery. Four of these satellites – Naiad, Thalassa, Despina and Galatea – actually orbit inside the ring system, within Neptune's Roche limit. At some point in the future these satellites, ranging from 60 to 160 km in diameter, may succumb to

► *Neptune and Triton, 16 arcseconds apart. This CCD image was taken by John Fletcher using a 250 mm SCT and CCD camera, September 8, 2004.*

gravitational tidal forces and be torn apart, forming a set of rings comparable in splendor to those of Saturn.

Triton, Neptune's largest satellite, was discovered by William Lassell using a 600 mm reflector, just 17 days after the discovery of Neptune itself. Triton orbits Neptune in the opposite direction to the planet's rotation in a period of 5 days 21 hours. Its orbital plane is highly inclined to the planet's equator. These orbital properties suggest that Triton might once have been an independent planet which was somehow captured by Neptune's gravity.

With a diameter of 2700 km, Triton is one of the largest satellites in the Solar System. Voyager 2 returned spectacular close-up images of its surface; some regions appear smooth and featureless, while other parts resemble the skin of a cantaloupe melon. No sizable impact craters were seen, indicating that the surface is being replenished by internal activity. Triton is thought to have a large rocky core surrounded by layers of ice. Dark patches appear to be sites of local crustal melting. One of the greatest surprises of the entire Voyager program came with the discovery of active geysers on Triton, which shoot quantities of nitrogen gas up to 20 km above the satellite's surface. Triton's activity is surprising, considering its extreme coldness – it is just 38°C above absolute zero.

Nereid, discovered photographically in 1949, is an irregularly shaped satellite with a diameter of 340 km. It has the most eccentric orbit of all the satellites in the Solar System, approaching Neptune as close as 1,400,000 km and receding as far as 9,700,000 km over a period of 360 days.

Observing Neptune

At opposition, which takes place around 2 days later each year, Neptune shines at magnitude 7.8. It is too faint to be seen with the unaided eye, but can be found easily enough through binoculars. Neptune has a mean opposition diameter of 2.5 arcseconds, so small that its disk cannot be readily perceived through anything smaller than a 100 mm telescope at high magnifications, although its distinct blue hue is just noticeable through smaller telescopes. Although Neptune is a more dynamic world than Uranus, its small apparent diameter means that little in the way of atmospheric features can be discerned through even the largest amateur telescopes.

Satellites

Shining at magnitude 13, Triton is the only Neptunian moon amenable to direct telescopic observation. A 200 mm telescope is large enough to show it, but it is theoretically bright enough to be glimpsed through a

► *Neptune, observed through a 150 mm refractor × 188 on July 28, 2002, by the author.*

75 mm telescope, given good seeing conditions and the use of an occulting bar to mask the brighter disk of Neptune. When at greatest elongation from Neptune, Triton appears a maximum of some 16 arcseconds from the planet. Experienced CCD imagers find little difficulty in securing images of Triton.

Recording Neptune

Using steadily held binoculars, careful observational drawings of the starfield in which Neptune appears will enable its slow movement to be discerned over a period of a few weeks and months. Few amateur astronomers routinely observe Neptune, though they may pay a high-magnification visit to the planet now and again just to reacquaint themselves with it, and perhaps to step up to the challenge of spotting Triton. Most observers are content to draw the entire star field surrounding Neptune, rather than making a large (probably blank!) disk drawing. It is worth preparing for the observation by consulting an ephemeris or planetarium program in order to be aware of the planet's tilt and its orientation through the eyepiece. It is also worth making a note or printout of the position of Triton and the background stars bright enough to be observed through your telescope.

TABLE 29 OPPOSITIONS OF NEPTUNE 2006–2016

Date	RA	Declination	Constellation
2006 Aug 11	21h 24m	−15° 30′	Capricornus
2007 Aug 13	21h 32m	−14° 52′	Capricornus
2008 Aug 15	21h 41m	−14° 13′	Capricornus
2009 Aug 17	21h 50m	−13° 33′	Capricornus
2010 Aug 20	21h 59m	−12° 51′	Capricornus
2011 Aug 22	22h 07m	−12° 09′	Aquarius
2012 Aug 24	22h 16m	−11° 25′	Aquarius
2013 Aug 27	22h 24m	−10° 41′	Aquarius
2014 Aug 29	22h 33m	−09° 56′	Aquarius
2015 Sep 01	22h 41m	−09° 10′	Aquarius
2016 Sep 02	22h 50m	−08° 23′	Aquarius

11 · PLUTO

Distant planetary outpost

Pluto, with a diameter of 2274 km, is the smallest of the nine planets; smaller, in fact, than seven planetary satellites, including our own Moon. Comparatively little is known about this distant world, the only major planet not to have been imaged at close range by a spaceprobe.

Searches for a planet beyond Neptune were begun in the early 20th century after some astronomers claimed that such an object was responsible for causing perturbations in the orbit of Neptune – just as Uranus' orbit is perturbed by Neptune itself. The photographic search for the ninth planet began at Lowell Observatory in 1927, and success finally came when Pluto's faint image was discovered on February 18, 1930, on photographic plates taken by Clyde Tombaugh using a 330 mm astrographic telescope. It soon became evident that Pluto was far too small to be the planet responsible for perturbing Neptune; however, we now know that there are no such perturbations, so the discovery of Pluto turns out to be a happy coincidence.

Pluto's large satellite, Charon, was discovered in 1978; measuring 1172 km across, it is more than half the diameter of Pluto. Both Pluto and Charon orbit their common center of gravity (around 1500 km outside of Pluto) in a period of 6.4 days, the two bodies being separated by an average distance of 19,640 km. Pluto and Charon have markedly different densities, and therefore different compositions. Lightweight Charon is likely to be composed chiefly of ices, while denser Pluto is thought to be made up of more than 50% rock mixed with ices. Such big differences in composition hint that the pair formed independently of each other; one theory suggests that Charon was formed as a result of a collision between Pluto and a small planet, similar to the way in which our own Moon is thought to have formed.

PLUTO: DATA	
Globe	
Diameter	2274 km
Density	2.0 g/cm^3
Mass (Earth = 1)	0.0021
Volume (Earth = 1)	0.0058
Sidereal period of axial rotation (retrograde)	6.387d
Escape velocity	1.1 km/s
Surface gravity (Earth = 1)	0.03
Albedo	0.9
Inclination of equator to orbit	122° 27′
Surface temperature (average)	45 K
Orbit	
Semimajor axis	39.54 AU = 5914 × 10^6 km
Eccentricity	0.249
Inclination to ecliptic	17° 06′
Sidereal period of revolution	248.54y
Mean orbital velocity	4.74 km/s
Satellites	1

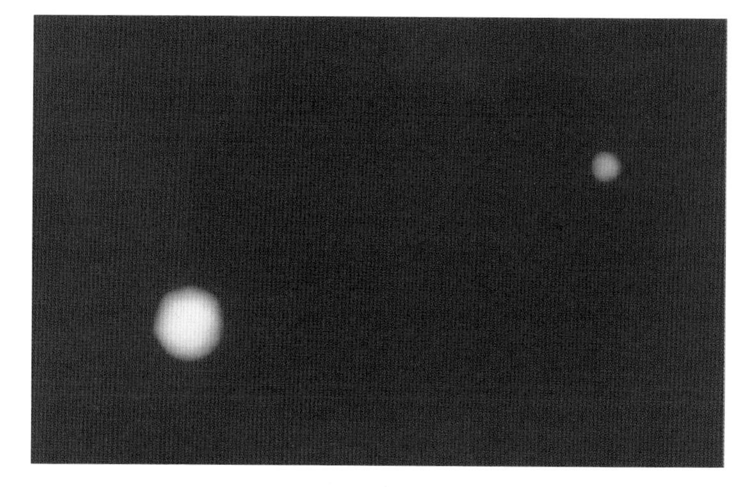

▲ *Pluto and Charon's relative sizes are evident in this Hubble image of February 1994.*

Orbit and rotation

Pluto circles the Sun in a period of 248.5 years in the most eccentric and highly inclined orbit of any of the nine major planets. It ranges from a minimum distance of 4.4 billion km to 7.5 billion km from the Sun, the latter figure being more than 50 times the distance between the Earth and Sun. For much of its orbit, Pluto is the most distant planet from the Sun, but for a period of around a decade on either side of its perihelion it is closer to the Sun than is Neptune. Pluto was nearer to the Sun than Neptune between 1979 and 1999; it shall be again between the years 2216 and 2236. Viewed in plan from above, it appears that the orbits of Pluto and Neptune actually cross each other; however, since the two planets' orbital planes are tilted to each other, there is no danger of a planetary collision at any point in the future.

Pluto takes 6 days 9 hours 18 minutes to revolve once on its axis; this is exactly the same as the orbital period of its satellite

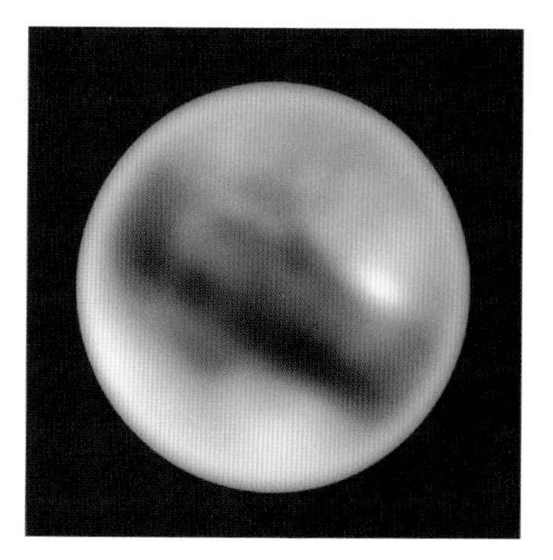

▶ *Pluto, imaged by the Hubble Space Telescope. The planet's surface displays a great range of tones.*

▲ *Pluto, imaged with a CCD camera at the prime focus of the 250 mm SCT (Meade LX200) of the Mount Tuffley Observatory, by amateur astronomer John Fletcher. Pluto was in the constellation of Ophiuchus, shining at a dim magnitude 14 when the image was captured on August 27, 2001.*

Charon – the two are tidally locked to each other. Seen from the surface of Pluto, Charon hovers at a constant height above the horizon, presenting the same face towards it; Pluto does the same in Charon's skies.

Pluto's polar axis is almost in line with the plane of its orbit around the Sun. By good fortune, the polar axis was at right angles to our line of sight between 1985 and 1990, giving rise to a series of mutual eclipses and occultations between Pluto and Charon which could be observed every day. With the best equipment available at the time, Pluto's disk was far too small to reveal any details, but analyses of the brightness of the light fluctuations enabled the construction of maps showing the albedo of the planet's surface. Pluto was found to have the biggest range in the reflectivity of its surface features of any planet beside the Earth. It was shown to have a brilliant south polar cap, a bright north polar cap, and a number of light and dark features in its equatorial region. This appearance was confirmed on images secured with the Hubble Space Telescope in 1993. Bright patches in the equatorial region may be impact craters surrounded by bright rays, and dusky circular regions may be large impact craters which have been filled with material exuded from beneath the planet's crust.

Observing Pluto

Pluto's orbital inclination of 17.1° to the ecliptic allows it to roam across an extensive swathe of the heavens. Indeed, Pluto can appear as far as 22°N or S of the celestial equator, appearing in constellations unfrequented by any other major planet, including Andromeda, Perseus, Triangulum, Microscopium and Sculptor.

At its maximum opposition brightness, Pluto shines at a feeble magnitude 13.6, so that it is theoretically possible to see it through a 150 mm telescope under good conditions. A 300 mm telescope will reveal Pluto under average conditions, but since it has a maximum apparent diameter of just 0.11 arcseconds it appears star-like even at the highest usable magnifications. Its disk is too small to be resolved visually through any amateur telescope.

For 10 years or so from 2006, Pluto winds a slow path across a rich section of the Milky Way, from the starfields of southern Serpens and into northwestern Sagittarius. Set against the star-studded Galactic plane, faint Pluto may prove harder than usual to pinpoint. To locate the planet, detailed charts showing stars down to around a magnitude higher than the faintest stars revealed by the observer's instrument are required. Such charts are readily obtainable from the observing sections of various astronomical societies, from numerous Internet sources, and are easily enough produced on all good planetarium programs.

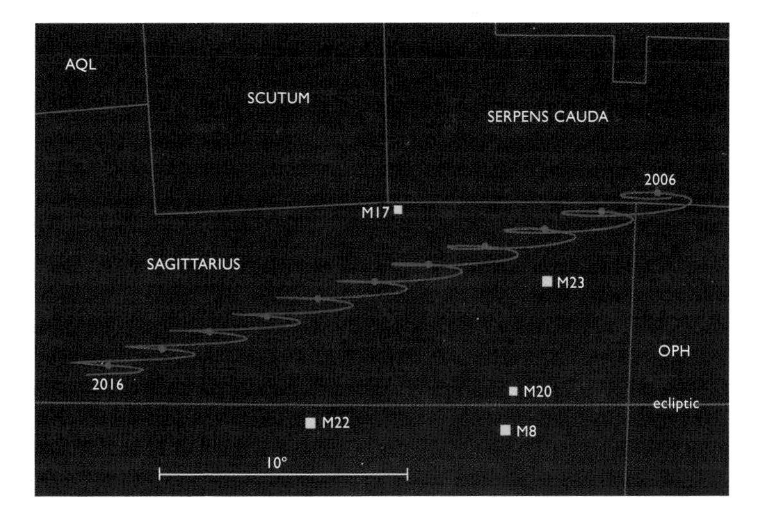

▲ *Pluto's path (red) among the stars of Serpens and Sagittarius, from 2006 to 2016. The locations of opposition are shown by the dots.*

TABLE 30 OPPOSITIONS OF PLUTO 2006–2016			
Date	RA	Dec	Constellation
2006 Jun 16	17h 41m	−15° 42′	Serpens Cauda
2007 Jun 19	17h 50m	−16° 23′	Sagittarius
2008 Jun 20	17h 59m	−17° 02′	Sagittarius
2009 Jun 23	18h 08m	−17° 39′	Sagittarius
2010 Jun 25	18h 17m	−18° 15′	Sagittarius
2011 Jun 28	18h 26m	−18° 48′	Sagittarius
2012 Jun 29	18h 35m	−19° 19′	Sagittarius
2013 Jul 02	18h 44m	−19° 49′	Sagittarius
2014 Jul 04	18h 52m	−20° 16′	Sagittarius
2015 Jul 06	19h 01m	−20° 41′	Sagittarius
2016 Jul 07	19h 10m	−21° 05′	Sagittarius

Like all superior planets, Pluto follows a retrograde path on either side of opposition, between quadrature; near quadrature the planet's motion slows to a virtual halt. Pluto moves at an angular velocity of more than one arcminute per day around opposition, so accurate drawings of the starfield in which Pluto is suspected to reside, made at intervals of a few days, are sufficient to show which of those faint stars is in fact the planet. It is exceedingly difficult, although not impossible, to detect Pluto's apparent motion during the course of a single evening. If Pluto is particularly close to a background star, or if it makes a pattern with two or more background stars, then accurate drawings made several hours apart may uncover such a slight movement.

Worlds beyond: Trans-Neptunian Objects

It is possible for the dedicated, well-equipped amateur to secure CCD images of a number of the brighter minor (and not so minor) planets orbiting the Sun beyond Neptune and Pluto. More than 1000 such Trans-Neptunian Objects (TNOs) have been discovered in recent years – 27 of them are larger than 500 km in diameter, while a few are in excess of 1000 km across. One of them, officially designated 2003 UB313, is substantially bigger than Pluto; unofficially known as "Xena," it might reasonably be considered the Solar System's tenth major planet. 2003 UB313 is, however, currently at a distance around twice that of Pluto and shines a rather dim 19th magnitude. It is far beyond visual location, but is bright enough to register on a CCD image taken through a standard astronomical CCD camera attached to a 200 mm telescope.

Dirty snowballs

Cometary nuclei are large solid agglomerations of silicate rock and ices – mainly water ice, but also various proportions of ammonia, methane, nitrogen, carbon monoxide and carbon dioxide ice. This composition has led to their somewhat ignominious description as "dirty snowballs." Most comets are thought to have formed out of the original interstellar material which condensed to form the Sun and the Solar System, although a few may have been gravitationally purloined from other nearby star systems. Comets therefore provide essential clues as to the composition of the original cloud from which the Solar System evolved.

Typically measuring just a few kilometers across, cometary nuclei are too small and faint to be visible through professional telescopes when they are much further away than the orbit of Neptune. If a cometary nucleus approaches the inner reaches of the Solar System, it begins to respond to the increased levels of solar heat and energy. At around the distance of Saturn, some 10 AU from the Sun, the heat received is sufficient to cause the most volatile ices on the surface – carbon monoxide and carbon dioxide – to sublimate, that is, to change from their solid state to a vapor without passing through an intermediate liquid phase. Water ice begins to sublimate at a distance of around 3 AU, which is about the distance of the main asteroid belt between Mars and Jupiter.

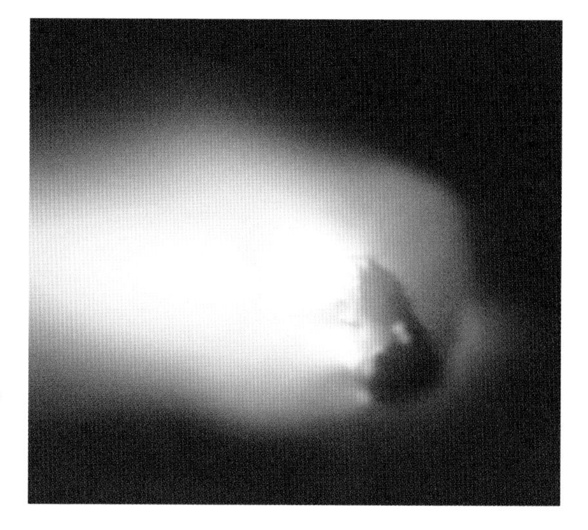

▶ Our first close-up view of a comet's nucleus came in 1986 when the Giotto spaceprobe passed within a few thousand kilometers of the nucleus of Halley's Comet.

▲ *Close-up of Comet Hale–Bopp, imaged by Robin Scagell in March 1997, showing its brilliant coma, and its gas and dust tails.*

Gases released from the surface of the comet's nucleus are bombarded by the solar wind, a 500 km/s stream of electrons and protons continually expanding away from the Sun into the depths of interplanetary space. The comet's gases ionize as they pick up electrical charge from the solar wind, and these ionized gases form a large cloud, known as a coma, around the nucleus. The coma can grow to as much as 100,000 km in diameter. Pressure from the solar wind pushes the ionized gases directly away from the Sun, and a long, straight ion tail is formed. This is known as a Type I tail.

The ices do not sublimate evenly across the nucleus' sunlit face; instead they sublimate most vigorously in certain locations around the coma, giving rise to jets. These streams of gas throw out small dust grains which were originally embedded in the icy crust. Pressure from the solar wind causes the released dust grains to lag behind the nucleus, producing a separate dust tail – known as a Type II tail – which develops alongside the gaseous ion tail. These dust particles provide the stuff of meteor streams.

Comets vary enormously in the amount of dust and gas that they produce. While some display just a broad fuzzy globular coma, others develop a magnificently structured coma, plus bright ion and dust tails. Some of the more active comets become really bright objects, and if conditions are right they can be impressive celestial spectacles for a period of time. Nobody who observed the glorious Comet Hale–Bopp during 1997 is likely to forget what a magnificent spectacle it made, with its bright star-like nuclear condensation, curved yellow dust tail and straight bluish ion tail. Other comets – indeed, the majority of comets observed – never reach naked-eye visibility and remain dim objects visible only through binoculars and telescopes.

A vast, hidden cometary realm

New comets are observed to fall into the Solar System from virtually any direction. No comet has been observed to have an orbit that suggests an interstellar origin, and there is a marked tendency for long-period cometary orbits to originate at a distance of about 50,000 AU from the Sun. Their place of origin far beyond the planets, in realms so distant that the Sun appears as a bright star-like point, is a zone populated by perhaps as many as six trillion cometary nuclei. Marking the very outer fringes of the Sun's gravitational domain, and stretching half way to the nearest stars, it is known as the Oort Cloud, named for Jan Oort, a Dutch astronomer who first postulated its existence in the mid-20th century.

The Oort Cloud is the source of all long-period comets. Orbiting the Sun in extremely elliptical orbits, these comets spend several millions of years as inert, deep-frozen chunks of material, only to indulge in a flurry of activity for a few months as they zip around the Sun before disappearing back into the depths of space. Their orbits can be drastically altered by close encounters with planets – so much so that they may be nudged into orbits with periods ranging from a few years to a few thousands of years. These comets, whose orbits are known in their entirety, are called periodic comets. As comets approach the Sun, jets of gas and dust emanating from their icy nuclei act like thrusters, modifying their orbits further – this introduces a slight and varying degree of unpredictability in the orbit of any comet.

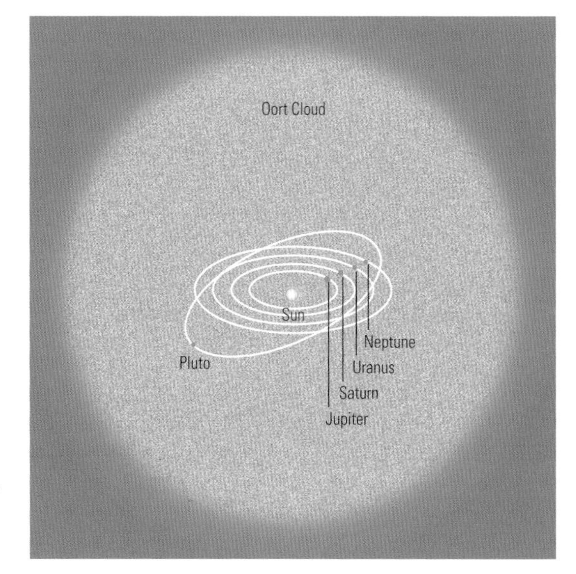

► *Far beyond the realm of the planets, trillions of cometary nuclei orbit the Sun within the Oort Cloud. These deep-frozen chunks of rock and ice occasionally visit the inner Solar System. (Illustration not to scale)*

Periodic comets

Periodic comets have well-known orbits lying within the realm of the planets, and they make regular, predictable visits to the inner Solar System. Most of them are too faint to observe through average backyard telescopes even when at their brightest near perihelion, but a number of them are familiar to amateur astronomers. Periodic comet Halley, for example, has a 76-year orbital period and an aphelion of some 35 AU (more distant than the orbit of Neptune). Orbiting the Sun every 3.3 years, periodic comet Encke has the shortest-known period of any comet, its orbital path taking it from aphelion inside the orbit of Jupiter to perihelion inside Mercury's orbit.

▲ Encke's Comet, a regular visitor to the inner Solar System. Observation by the author on November 28, 2003, at 18:35 using 15 × 70 binoculars.

Named for their discoverer (or discoverer of their periodicity), a total of 164 periodic comets are currently designated a number on a list maintained by the Minor Planet Center. They are prefixed with P (for periodic), and listed in order of the discovery of their periodic nature. Thus, Halley's Comet is referred to as 1P Halley, Encke's Comet as 2P Encke, and so on. In addition, there are more than 160 periodic comets without a designated number and several dozen asteroids within the Solar System which display all the attributes of cometary nuclei.

Many periodic comets are thought to have originated in the Oort Cloud, their orbital paths having been modified to far shorter periods after being diverted by close encounters with the planets, chief among them the giant planet Jupiter. It is likely that some of the smaller satellites in odd orbits around Jupiter, Saturn, Neptune and Pluto are actually captured comets.

Cometary nomenclature

If a new comet is discovered visually, or first captured on a photograph or CCD image secured by an amateur astronomer, the comet is named in that observer's honor. Occasionally, multiple observers share the honor of a comet's discovery, the order of naming reflecting the date or time of each codiscoverer's observation. If the comet is discovered on a CCD image made during one of the professional automated comet/asteroid searches, or on an image taken by a satellite, the comet is usually named for the name of the search program or the satellite. For example, the many recently discovered LINEAR comets are named for the Lincoln Near Earth Asteroid Research (LINEAR) project, a MIT Lincoln Laboratory program funded by the US Air Force. Similarly, there are comets named for NEAT (Near-Earth Asteroid Tracking), a program run by NASA and the Jet Propulsion Laboratory, to discover near-Earth objects and LONEOS (the Lowell Observatory Near-Earth Object Search). Comets have also been named for the Infrared Astronomical Satellite (IRAS) and the Solar and Heliospheric Observatory (SOHO).

New comets are designated with the letter C, followed by the year of discovery, a letter indicating the half-month of the date of its discovery, a number indicating its order of discovery in that half-month, and finally the discoverer's name(s).

The letters denoting a comet's half-month of discovery are listed in Table 31.

TABLE 31 COMETARY DESIGNATIONS

A	Jan 1–15	B	Jan 16–31	C	Feb 1–15	D	Feb 16–29
E	Mar 1–15	F	Mar 16–31	G	Apr 1–15	H	Apr 16–30
J	May 1–15	K	May 16–31	L	Jun 1–15	M	Jun 16–30
N	Jul 1–15	O	Jul 16–31	P	Aug 1–15	Q	Aug 16–31
R	Sep 1–15	S	Sep 16–30	T	Oct 1–15	U	Oct 16–31
V	Nov 1–15	W	Nov 16–30	X	Dec 1–15	Y	Dec 16–31

Since the letter I may be confused with the number 1, it is not used, and neither is there need for a letter Z in the scheme.

For example, Comet Hale–Bopp, independently discovered on July 23, 1995, by amateur astronomers Alan Hale in New Mexico and Thomas Bopp in Arizona, was the first comet discovered in the latter half of July 1995. It therefore assumed the official designation of C/1995 O1 Hale–Bopp. The 1995 tag appeared to confuse some observers viewing the comet in 1997, unaware that it had been discovered two years earlier when it was beyond Jupiter's orbit, at a distance of more than 7 AU – the furthest comet ever discovered by amateurs.

Comets which have disappeared or no longer exist are designated with the prefix letter "D," while comets whose orbits cannot be computed with any degree of reliability due to a lack of precise observations take the prefix "X."

Great comets

Only under exceptional circumstances can any of the periodic comets appear bright enough to be easily seen with the unaided eye, and they rarely assume spectacular proportions in the night skies. A relatively small, intrinsically faint comet can appear bright and impressive in the night sky if its orbit brings it into close proximity with the Earth; a far bigger, intrinsically brighter comet can appear small and faint if its perihelion takes place on the opposite side of the Sun to the Earth. Optimum circumstances for any comet occur when it is at its intrinsically brightest near perihelion and in close proximity to the Earth at the same time, with a sizable apparent angular distance from the Sun in order that it is observed against a dark sky.

Halley's Comet, for example, appeared pretty unimpressive to the non-astronomer during its 1985–86 apparition; when at its brightest near perihelion in early February 1986, the comet was on the opposite side of the Sun and was not visible. It appeared considerably more spectacular back in 1910, when its perihelion passage was observable. Then, Halley's Comet was an easy naked-eye object, sporting a bright tail. However, it was outshone by an even brighter new comet, fresh from the Oort Cloud, known as the Great January Comet of 1910; this impressive object was actually visible during the daytime around the time of its perihelion. We have to go back in history nearly 1200 years to find the most spectacular apparition of Halley's Comet – in April AD 837 the comet approached the Earth to within 5.1 million km, developed a tail around 60° long and was marveled at by people around the world.

Most newly discovered comets never develop into anything more spectacular than dim objects that can only be faintly glimpsed through binoculars or telescopes. Now and again – say, once or twice a decade – a comet bright enough to be viewed with the unaided eye becomes visible. Really spectacular bright comets are rare, with only a handful coming into view every century. There have been 26 truly magnificent comets visible since the invention of the telescope.

▲ *Halley's Comet near the Pleiades, observed by the author through 7 × 50 binoculars on November 17, 1985, at 00:35 UT.*

Most of the readers of this book will have been lucky enough to have viewed two of these – Comet Hyakutake in 1996 and Comet Hale–Bopp in 1997 – but doubtless there are many readers who are old enough to remember seeing Comet Ikeya–Seki of 1965, Comet Bennett of 1969 and Comet West of 1975. As of the time of writing, the 21st century has yet to see a Great Comet.

TABLE 32 BRIGHT COMETS OF THE TELESCOPIC ERA
C/1618 W1
C/1664 W1
C/1665 F1
C/1668 E1
C/1680 V1
1P/Halley (1682)
C/1686 R1
C/1743 X1
C/1769 P1 Messier
Great Comet C/1807 R1
Great Comet C/1811 F1
Great March Comet C/1843 D1
C/1858 L1 Donati
Great Comet C/1861 J1
Great Southern Comet C/1865 B1
C/1874 H1 Coggia
Great September Comet C/1882 R1
Great Comet C/1901 G1
Great January Comet C/1910 A1
1P/Halley (1910)
C/1927 X1 Skjellerup–Maristany
C/1965 S1 Ikeya–Seki
C/1969 Y1 Bennett
C/1975 V1 West
C/1996 B2 Hyakutake
C/1995 O1 Hale–Bopp

Astronomers find it notoriously difficult to predict accurately the brightness curve of a newly discovered comet. Of course, it's a safe bet to assume that a comet will brighten and its coma will enlarge as it nears the Sun. The proximity and position of the Earth with respect to a comet play a great part in its apparent brightness. A comet with a big, bright, active nucleus, observed when it is near perihelion against a dark background sky, is likely to be an impressive sight, perhaps justifying the title of "Great Comet." Comet Kohoutek of 1973 was discovered as a fairly bright object while in deep space, and all the indications were that it was likely to become the "Comet of the Century" – yet Kohoutek dismally failed to live up to expectations and proved to be a great disappointment.

Observing comets

Locating comets

Ephemerides of currently visible or soon-to-be-visible comets are featured on various websites and in a variety of astronomical publications. A highly recommended website is JPL's Comet Observation Home Page at http://encke.jpl.nasa.gov/.

Ephemerides give the predicted position of a comet's nucleus in terms of its Right Ascension (RA) and Declination (Dec) in set intervals of time (usually at 00h UT on each given date). Most ephemerides also give some indication of how bright the comet is expected to be on each date, based on early observations and typical cometary light curves.

Once a comet's predicted position is known, its path can be plotted in pencil upon a suitable star chart, with small tick marks and labels at intervals giving the date and time of its predicted position. My own old and well-used copy of *Norton's Star Atlas*, for example, has at least one comet's path marked on each page, including such notable visitors as Iras–Araki–Alcock of 1983, Halley's Comet, Brorsen–Metcalf of 1989, Swift–Tuttle of 1992 – in fact, virtually every reasonably bright comet visible from my location from the 1980s to the mid-1990s. Once a path is plotted, the observer is free to interpolate a comet's position from the available data.

As more positional measurements of a newly discovered comet are made, its orbital parameters can be refined and made more reliable. It is common to see the RA and Dec figures given in an initial ephemeris differing slightly from those in an updated ephemeris. Additionally, the accuracy of ephemeris predictions may be further thrown out as a result of activity on the comet's nucleus: jets of material thrown out by the nucleus act like rocket thrusters and are capable of modifying the orbit of a comet in unpredictable ways.

Computer programs

Many of the better astronomical programs for personal computers are capable of displaying the positions of known periodic comets with a high degree of accuracy, in addition to giving information about the comet's predicted magnitude and other data. It is possible to download data on the orbital parameters of newly discovered comets and input them into a variety of planetarium programs for PC.

The website http://cfa-www.harvard.edu/iau/Ephemerides/Comets/SoftwareComets.html gives the orbital elements of observable comets in formats suitable for loading into a number of popular planetarium-type software packages. Some programs, such as Starry Night Pro, have the facility of automatically updating their data on connection to the Internet, ensuring that the latest ephemerides are available.

While computer programs are a great observational and research tool, they are only as good as the data available to them. Although they will show a comet according to its predicted brightness, displaying a nice graphic image of the comet and the general direction of its tail, they are not sophisticated enough to be able to predict the actual behavior of a comet – the graphics that they display generally fail to show what the object will actually look like through binoculars or the telescope eyepiece.

Computerized telescopes

Meade and Celestron go-to telescopes are provided with electronic handsets which have a built-in library of ephemerides for any number of periodic comets. At the touch of a few buttons it is possible to have your telescope slew around to, say, the latest position of Halley's Comet (though it will certainly be invisible) and have a list of relevant data about the comet scroll across the handset display. Additionally, it is possible to input data for a new comet or update the handset's software via a cable attached to a PC with Internet access. It is also possible to input a specific RA and Dec from an ephemeris and have the telescope slew to that position. If the comet is not visible, it is either too faint to see or it is not within the field of view. In the latter case, the computerized telescope can be instructed to perform a spiral search around the position it is currently pointing toward. All of these facilities are certainly very handy.

Visual detail within comets

Bright comets can be enjoyed with the unaided eye. However, light pollution is a great handicap when attempting to discern any of the subtle detail within a comet's tail(s). For example, from the suburbs of a major city, the bright magnitude 0 head of Comet Hyakutake of 1996 could easily be seen without optical aid, but little else. From a dark countryside location, the comet was splendid with the unaided eye, with a narrow tail some 70° long.

Binoculars, with their wide field of view, are ideal for general comet observing, though even they may not have a field of view wide enough to take in all of a bright comet's tail. Binoculars are also better at delivering color to the eyes than a telescope eyepiece.

Telescopes are capable of being used at fairly high magnifications to view detail in and around the nuclear regions of a bright comet; the more magnification that is used, the amount of faint detail visible decreases, but structure within the brightest parts of the nuclear region becomes easier to discern.

▲ *A close-up view of the detail in and around Comet Hale–Bopp's coma, seen on March 20, 1997, at 04:50 UT. Observational drawing by the author, using a 150 mm Newtonian.*

False nucleus

At the very heart of the comet lies its nucleus – a "dirty snowball" ranging in size from a few kilometers to several tens of kilometers. Cometary nuclei themselves are actually rather dark objects, and are to all intents and purposes unobservable through amateur telescopes. Instead, the tiny bright spot observed at the center of the coma is known as the "false nucleus," its brightness being produced by the concentration of gases and dust being ejected from the real nucleus' surface. Although some comets (especially those at some distance from the Sun) may appear as nebulous patches, with little obvious structure besides a certain brightening toward an offset brighter central area, given good observing conditions and a sufficiently large telescope, most comets display a false nucleus. It may appear as a faint, star-like point in dimmer comets, while a large, bright active comet may have a false nucleus with an apparent angular diameter of several arcseconds.

Jets and shells

As the comet approaches perihelion, its most frenetic phase begins. In large active comets the false nucleus is often noticeably asymmetric in shape and may extend into bright streamers. These features are produced by active jets on the surface of the nucleus; as the nucleus rotates, the streamers assume a curved or swirling pattern. At the same time, areas with differing levels of potential activity are brought into the sunlight, producing variations in the brightness and shape of the false nucleus over a period of a few hours. Material puffed out by active areas on the nucleus can form a series of distinct expanding layers of material; these may be clearly visible on the sunward side of the nucleus, each curving away from the Sun with the bow shock of the solar wind.

Tails

Most comets develop both a Type I tail of ionized gas and a Type II tail of dust, although their real and apparent size and form vary enormously depending on the comet and the angle at which we view it. Comet tails can grow to enormous lengths of millions, tens of millions or (rarely) hundreds of millions of kilometers.

Type I tails are straight and point directly away from the Sun, regardless of the actual motion of the comet through space; seemingly counterintuitively, a comet races in the general direction of its ion tail after perihelion. Binocular observation may reveal a considerable amount of detail within the ion tail, including wave-like patterns and knots of brighter material. Type I tails can take on a distinctly blue color – evident with the unaided eye in the ion tails of Comet Hyakutake and Comet Hale–Bopp – and may be particularly striking through binoculars.

Composed of myriad tiny dust particles released from the nucleus and shining by reflected sunlight, the Type II dust tail of a bright comet often assumes a yellow or orange-brown hue. Each particle released by the nucleus continues along the same path as the comet, but solar radiation pressure pushes them away from the Sun. Forced to lag behind, the particles create a dust tail that appears distinctly curved.

Antitails, such as the famous "spike" which appeared to protrude from the head of C/1956 R1 Arend–Roland, are simply line-of-sight effects, produced when the Earth is level with the comet's orbital plane and distant parts of the curving dust tail appear to precede the comet's head.

▲ *Comet Ikeya–Zhang, observed by the author on May 3, 2002, at 21:30 UT using a 250 mm Newtonian.*

Recording comets

It is tempting to regard pencil drawings as a quaint and somewhat outmoded means of recording the appearance of any celestial object. In the case of comet observing, however, a drawing of what can be seen with the eye through the telescope eyepiece can deliver more data than a CCD image. The eye has a very wide dynamic range, capable of taking in subtleties of tone as well as intricate detail. CCD cameras are not good at conveying subtle variations in brightness. While viewing a comet through the telescope eyepiece, be on the look out for the shape of the false nucleus and any jets emanating from it, as well as subtle detail within the coma. Using steadily held binoculars, look out for waves, condensations and bright knots in the comet's tail(s), and note their shape, orientation, length and any subtle coloration that may be visible.

Drawing

The drawing itself is best made on a predrawn circular blank of smooth white paper using a soft pencil (dark graphite, such as 3B or below). First, mark the position of the bright nuclear condensation; if the whole comet can be comfortably seen within the field of view, place the comet at the center. If the comet is bright and extends out of the field, place the comet's nucleus in a position where most overall detail can be seen. Next, draw in the brightest stars within the field of view as accurately as possible; if it's possible to identify any of them, note their name. Once the main stars have been sketched, draw in the fainter stars. Returning to the comet, sketch the approximate size and shape of the nuclear brightening. Using layers of lightly applied pencil, draw in the coma and any detail visible within the tail(s). The tip of your index finger serves nicely as a smudger, blending in the pencil where required, and a sharpened eraser can be carefully used to denote any clear-cut regions of darkness within the coma or tail. An alternative to making a "negative" pencil drawing is to use pastels on dark paper; most prefer the convenience of ordinary pencil.

▲ *Comet Ikeya–Zhang passes around 2° west of globular cluster M13 in Hercules. Observation by the author on May 16, 2002, at 01:50 UT, using 7 × 50 binoculars.*

In addition to the notes that ought to accompany every observational drawing – date, time (in UT), instrument, seeing, and so on – make a written note to clarify any interesting or unusual features. It is also important to note the orientation of the field of view depicted in your drawing, marking celestial north, south, east and west around the edge of the drawing – special care must be taken to get this right if you're observing with a star diagonal. Another useful indication is to mark the drift direction of features across the field of view with an arrow at the edge of the circle.

Photography

Bright comets of naked-eye magnitude or brighter can be reasonably well imaged using a regular 35 mm compact or SLR camera, unguided, from a dark sky site. The amount of detail you'll capture depends on the exposure, but a few minutes' exposure will be enough to capture all of the naked-eye stars as well as the coma and perhaps hints of a tail, all without much trailing. The CCD chips within high-end digital cameras are more sensitive than photographic film, and they are quite capable of capturing the appearance of a bright comet in an exposure of a few tens of seconds.

A 35 mm compact or SLR camera piggybacked to a driven telescope, or mounted on its own equatorially driven platform, will enable much longer exposures to be made, allowing fainter background stars to be seen as well as revealing more color and detail within the coma and tail(s). If the platform is accurately aligned and driven, an SLR can be fitted with a telephoto lens for a closer view of the coma and detail within it.

The best results are obtained by using an equatorially driven telescope as a large telephoto lens, the SLR being used at the telescope's prime focus (with the telescope minus its eyepiece and the camera minus its lens). Since a comet is in motion against the background stars, the best results during long exposures at relatively high magnifications will be obtained if the comet is kept at the center of the field of view. This can be achieved using a guide telescope or a flip mirror

CCD imaging

A regular modified webcam is perfectly capable of securing detailed images of the nuclear region of a bright comet, while an astronomical CCD camera capable of time exposures will enable advanced studies to be made. CCD imaging opens up a host of possibilities, including recording comets too faint to be viewed with the unaided eye, capturing detail within the coma of bright comets (including jets and shells), recording transient brightness fluctuations of the nucleus and making advanced astrometric measurements of a comet's position in the sky.

◀ Comet Machholz near the Pleiades, imaged on January 4, 2005, by Pete Lawrence.

▼ The bright coma of Comet Machholz, imaged by Dr. Richard Prettyman on January 30, 2005, at 20:55 UT, using a. 250 mm Newtonian.

13 · METEORS

Meteors

A meteor – commonly called a "shooting star" – is the glowing trail produced when a small particle of dust, known as a meteoroid, is heated by friction with the atmosphere and burns up at a height between around 50 and 200 km. The appearance of each meteor is sudden and unpredictable, and each meteor flash is visible for only a fraction of a second as it crosses a portion of the night sky. Generally speaking, the larger the meteoroid, the brighter the meteor it produces. A faster meteoroid burns up higher in the atmosphere, since it collides with more atmospheric molecules per second than its slower counterpart. A meteor's apparent brightness varies with its distance from the observer, its elevation above the observer's horizon and atmospheric clarity.

Meteors come from two sources – cometary trails and general interplanetary debris. Comets leave trails of dust particles in their wake, and these particles are spread along the comet's orbit. Throughout the year, on certain predictable dates, the Earth passes through dozens of these trails, producing annual meteor showers, some of which regularly feature particularly bright meteors known as fireballs. Interplanetary debris burning up in the Earth's atmosphere produces sporadic meteors and fireballs; some of the material is so large and substantial that it survives its superheated passage through the atmosphere and impacts on the Earth's surface as a meteorite. Most meteorites originate from asteroids, but a rare few come from the Moon and Mars; none of them are associated with the annual meteor showers.

Zodiacal light and the gegenschein

Countless trillions of dust particles – material left over from the formation of the Solar System, debris from asteroid collisions and the like – orbit the Sun in a broad flattened disk roughly aligned with the plane of the ecliptic. Sunlight scattered from these particles enables the disk to be seen as a diffuse, faintly glowing cone extending from the western horizon after evening twilight or the eastern horizon before the start of morning twilight. Known as the zodiacal light, this phenomenon is notoriously difficult to see from anywhere but a site with clear, transparent skies because of its low surface brightness. It is best seen on spring evenings or autumn mornings, when it makes its greatest angle with the horizon.

A related phenomenon, the gegenschein, can be seen as a faint glowing patch diametrically opposite the Sun; it is best placed at midnight

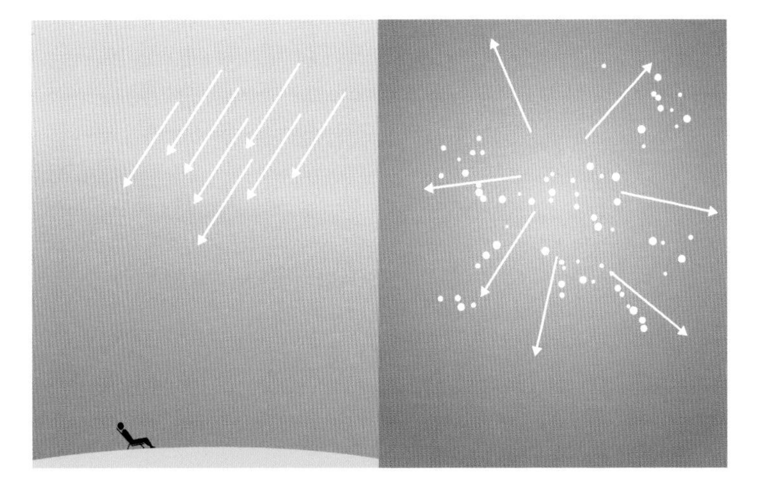

▲ *Meteors belonging to the same shower enter the atmosphere on parallel paths.* *From the observer's perspective, they appear to radiate from a single point.*

under exceptionally dark late winter skies when it is at its highest point above the horizon. The sight of these phenomena will impress upon the observer that interplanetary space is by no means empty.

Although photographs are capable of capturing the zodiacal light and the gegenschein, the phenomena are too faint and diffuse to be visible through binoculars or telescopes. The unaided eye is by far the best way of appreciating these elusive spectacles.

Annual meteor showers

Every year the Earth passes through a number of meteoroid streams, producing meteor showers that occur at around the same date each year. Meteors produced by the annual showers appear to come from a small, well-defined part of the sky known as the radiant. It is our perspective from the Earth, passing through the meteor stream, that produces the radiant effect. Each shower radiant is named for the constellation in which it lies, or the brightest nearby star. Each meteor shower has its own particular delights, and each is capable of springing surprises from year to year.

The zenithal hourly rate (ZHR) is a theoretical value given for the number of meteors a single observer would see if the radiant was directly overhead in a dark sky site with a limiting magnitude of 6.5 (the limit of naked-eye visibility under dark skies). These ideal circumstances rarely arise in combination, so the actual observed rates are usually much lower than the ZHR.

TABLE 33 MAJOR METEOR SHOWERS

Bold text indicates showers emanating from radiants in the north celestial hemisphere. ***Bold italic text*** indicates major showers described below. Plain text indicates southern hemisphere showers.

Shower	Activity span	Maximum	Radiant RA / Dec	Velocity km/s	ZHR
Quadrantids	***Jan 01–Jan 05***	***Jan 03***	***15h 20m / +49°***	***41***	***120***
Gamma Velids	Dec 24–Jan 23	Jan 05	08h 20m / −47°	35	2
Alpha Hydrids	Jan 05–Feb 14	Jan 14	08h 52m / −11°	44	2
Alpha Crucids	Jan 06–Jan 28	Jan 15	12h 48m / −63°	50	3
Alpha Carinids	Jan 24–Feb 09	Jan 30	06h 20m / −54°	25	2
Alpha Centaurids	Jan 28–Feb 21	Feb 07	14h 00m / −59°	56	6
Delta Velids	Jan 22–Feb 21	Feb 14	08h 44m / −52°	35	1
Omicron Centaurids	Jan 31–Feb 19	Feb 14	11h 48m / −56°	51	2
Theta Centaurids	Jan 23–Mar 12	Feb 14	14h 00m / −41°	60	4
Delta Leonids	**Feb 15–Mar 10**	**Feb 24**	**11h 12m / +16°**	**23**	**2**
Gamma Normids	Feb 25–Mar 22	Mar 13	16h 36m / −51°	56	8
Delta Pavonids	Mar 11–Apr 16	Mar 29	20h 30m / −63°	60	5
Lyrids	***Apr 16–Apr 25***	***Apr 22***	***18h 04m / +34°***	***49***	***18***
Pi Puppids	Apr 15–Apr 28	Apr 24	07h 20m / −45°	18	var
Eta Aquarids	***Apr 19–May 28***	***May 05***	***22h 32m / −01°***	***66***	***60***
Eta Lyrids	**May 5–May 17**	**May 09**	**19h 12m / +44°**	**44**	**3**
Beta Corona Australids	Apr 23–May 30	May 16	18h 56m / −40°	45	3
June Lyrids	**Jun 11–Jun 21**	**Jun 16**	**18h 32m / +35°**	**31**	**3**
Tau Cetids	Jun 18–Jul 04	Jun 27	01h 36m / −12°	66	4
June Boötids	**Jun 26–Jul 02**	**Jun 27**	**14h 56m / +48°**	**18**	**var**
Tau Aquarids	Jun 19–Jul 05	Jun 29	22h 48m / −12°	63	3
Pegasids	**Jul 07–Jul 13**	**Jul 09**	**22h 40m / +15°**	**70**	**3**
July Phoenicids	Jul 10–Jul 16	Jul 13	02h 08m / −48°	47	var
Alpha Cygnids	**Jul 11–Jul 30**	**Jul 18**	**20h 20m / +47°**	**37**	**3**
Piscis Austrinids	Jul 15–Aug 10	Jul 28	22h 44m / −16°	35	5
South Delta Aquarids	***Jul 12–Aug 19***	***Jul 28***	***22h 36m / −30°***	***41***	***20***
Alpha Capricornids	Jul 03–Aug 15	Jul 30	20h 28m / −10°	23	4
South Iota Aquarids	Jul 25–Aug 15	Aug 04	22h 16m / −15°	34	2
North Delta Aquarids	Jul 15–Aug 25	Aug 08	22h 20m / −05°	42	4
Perseids	***Jul 17–Aug 24***	***Aug 12***	***03h 04m / +58°***	***59***	***100***
Kappa Cygnids	**Aug 03–Aug 25**	**Aug 17**	**19h 04m / +59°**	**25**	**3**
North Iota Aquarids	Aug 11–Aug 31	Aug 19	21h 48m / −06°	31	3
Pi Eridanids	Aug 20–Sep 05	Aug 27	03h 28m / −15°	59	4
Gamma Doradids	Aug 19–Sep 06	Aug 30	04h 04m / −50°	41	5
Alpha Aurigids	**Aug 25–Sep 08**	**Sep 01**	**05h 36m / +42°**	**66**	**7**
September Perseids	**Sep 05–Sep 16**	**Sep 09**	**04h 00m / +47°**	**64**	**5**
Kappa Aquarids	Sep 08–Sep 30	Sep 18	22h 36m / −05°	19	3

Shower	Activity span	Maximum	Radiant RA / Dec	Velocity km/s	ZHR
Delta Aurigids	**Sep 16–Oct 10**	**Sep 23**	**05h 00m / +49°**	**64**	**3**
Giacobinids	**Oct 06–Oct 10**	**Oct 08**	**17h 28m / +54°**	**20**	**var**
Epsilon Geminids	**Oct 14–Oct 27**	**Oct 18**	**06h 48m / +27°**	**70**	**2**
Orionids	***Oct 02–Nov 07***	***Oct 21***	***06h 20m / +16°***	***66***	***23***
Leo Minorids	**Oct 23–Oct 25**	**Oct 24**	**10h 48m / +37°**	**61**	**2**
Southern Taurids	**Oct 01–Nov 25**	**Nov 05**	**03h 28m / +13°**	**27**	**5**
Northern Taurids	**Oct 01–Nov 25**	**Nov 12**	**03h 52m / +22°**	**29**	**5**
Delta Eridanids	Nov 06–Nov 29	Nov 12	03h 40m / −02°	31	1
Zeta Puppids	Nov 02–Dec 20	Nov 13	07h 48m / −42°	41	3
Leonids	***Nov 14–Nov 21***	***Nov 17***	***10h 12m / +22°***	***71***	***20+***
Alpha Monocerotids	**Nov 15–Nov 25**	**Nov 21**	**07h 48m / +01°**	**65**	**var**
Dec Phoenicids	Nov 28–Dec 09	Dec 06	01h 12m / −53°	18	var
Puppid/Velids	Dec 01–Dec 15	Dec 07	08h 12m / −45°	40	10
Monocerotids	**Nov 27–Dec 17**	**Dec 09**	**06h 40m / +08°**	**42**	**3**
Sigma Hydrids	**Dec 03–Dec 15**	**Dec 12**	**08h 28m / +02°**	**58**	**2**
Geminids	***Dec 07–Dec 17***	***Dec 14***	***07h 28m / +33°***	***35***	***120***
Coma Berenicids	**Dec 12–Jan 23**	**Dec 19**	**11h 40m / +25°**	**65**	**5**
Ursids	**Dec 17–Dec 26**	**Dec 22**	**14h 28m / +76°**	**33**	**10**

Major meteor showers

Quadrantids

Active: Jan 01–Jan 05
Maximum: Jan 03
Radiant: 15h 20m / +49°
Velocity: 41 km/s
Max ZHR: 120

Taking its name from the long-defunct constellation of Quadrans Muralis from which it was once seen to radiate, the year's first meteor shower makes its appearance on the first day of each new year. The faint stars of Quadrans lie within and to the northeast of the constellation of Boötes, on the border of Draco and Hercules. Quadrantid meteors are mostly faint, and shoot across the sky at a medium speed. A bluish hue has been noted in many of the brighter meteors. About 10% leave a glowing train.

Undoubtedly the finest recorded Quadrantid display took place in 1909, with an impressive ZHR of over 200 meteors per hour. In 1965 a rate of 190 was observed. In 1973 and 1985 there were ideal conditions for the Quadrantids, with the Earth passing through a dense part of the meteor stream, and high rates were again seen. Before the 19th century the Quadrantid stream did not intersect the Earth's orbit, and gravitational perturbations will eventually shift the stream out of our path again. The shower will have perhaps dwindled out of existence by the turn of the 22nd century. The 1909 event probably represented the Quadrantids' finest hour. No parent comet has been identified.

Lyrids

Active: Apr 16–Apr 25
Maximum: Apr 22
Radiant: 18h 04m / +34°
Velocity: 49 km/s
Max ZHR: 18

Lyrids are brilliant, medium-speed meteors, some leaving persistent trains. The Lyrid stream is associated with Comet Thatcher of 1861, a comet with a 415-year period. Records of Lyrid storms can be found in Chinese annals of 687 and 15 BC. The last true Lyrid storm was seen from the eastern USA in 1803, where a rate of 700 meteors per hour was recorded during a two-hour burst of activity. The Lyrids are well worth keeping a watch for, since bursts of activity – even a major storm – may be produced if the Earth passes through large concentrations of material.

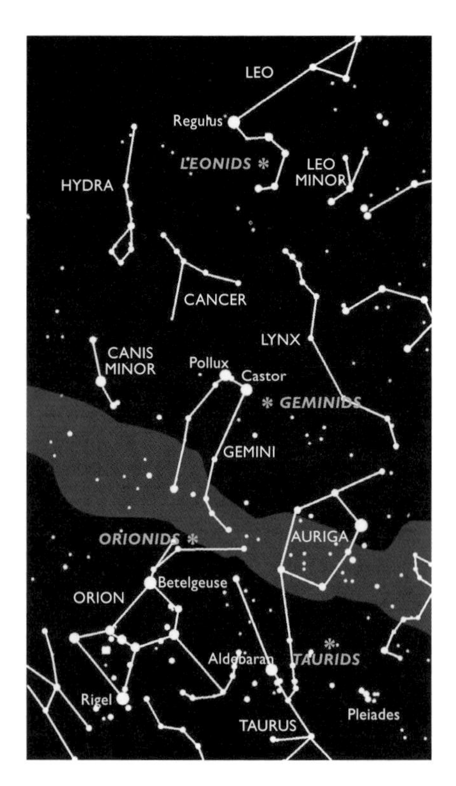

▲ *Radiant locations of several annual meteor showers which occur later in the year.*

Eta Aquarids

Active: Apr 19–May 28
Maximum: May 05
Radiant: 22h 32m / −01°
Velocity: 66 km/s
Max ZHR: 60

The Eta Aquarids are very fast meteors with persistent trains. They are part of the same meteor stream as the Orionids (detailed below), seen in October. This meteor stream is the only major one to be encountered by the Earth twice in a year. The Earth passes closer to the center of the stream during May than it does in November, so Eta Aquarid rates are higher than those of the Orionids.

South Delta Aquarids

Active: Jul 12–Aug 19
Maximum: Jul 28
Radiant: 22h 36m / −30°
Velocity: 41 km/s
Max ZHR: 20

Viewed from the UK or northern USA, this shower's radiant is at a low altitude, and observed rates suffer as a consequence. Most Delta Aquarids are faint, though there are sometimes brilliant examples. They travel in long paths at a medium speed. Around 10% of the brighter meteors leave persistent trains.

Perseids

Active: Jul 17–Aug 24
Maximum: Aug 12
Radiant: 03h 04m / +58°
Velocity: 59 km/s
Max ZHR: 100

The Perseid meteor shower is one of the most popular of the annual showers. Its meteors are very fast and often extremely bright, frequently flaring midflight and leaving glowing trains. Good rates can be expected a day or more on either side of the date of maximum. During its period of activity the Perseid radiant moves from beneath the "W" of Cassiopeia, though the double cluster of h and Chi Persei and into Camelopardalis. Perseid rates have often been high, with notable showers in 1980, 1981, 1991, 1992 and 1993.

The Perseid meteor stream is associated with Comet Swift–Tuttle. It is very old and is thought to have been around for perhaps more than 2000 years. The earliest record of the shower comes from China in AD 36, and the first European account was given in AD 811. The Perseid shower was once commonly known as the "burning tears of St Lawrence" because they peaked on August 10, around St Lawrence's feast day.

Orionids

Active: Oct 02–Nov 07
Maximum: Oct 21
Radiant: 06h 20m / +16°
Velocity: 66 km/s
Max ZHR: 23

Orionids are bright, very fast meteors, with some fireballs produced during each period of activity, particularly about 3 days after maximum. The Orionid stream, a product of Halley's Comet, is rather complicated. The radiant is multiple, though this is not obvious to the visual observer because the radiants are grouped together in the same small area of the sky. Activity fluctuates from year to year.

Leonids

Active: Nov 14–Nov 21
Maximum: Nov 17
Radiant: 10h 12m / +22°
Velocity: 71 km/s
Max ZHR: 20+

From its radiant in the sickle asterism of the head of Leo, the Leonid shower produces very fast meteors with persistent trains. They are the fastest produced by any of the annual showers. Enhanced Leonid activity occurs from time to time, with spectacular meteor storms taking place roughly every 33 years, when the Earth passes through denser parts of the Leonid stream. Comet Tempel–Tuttle is the known parent body, and it rounds the Sun every 33 years depositing large amounts of debris. More than a millennium of recorded Leonid activity has produced many spectacular Leonid showers and some incredible storms.

Geminids

Active: Dec 07–Dec 17
Maximum: Dec 14
Radiant: 07h 28m / +33°
Velocity: 35 km/s
Max ZHR: 120

The Geminid shower is currently the best of the annual meteor showers, producing high ZHRs of very bright, intensely white, slow-moving meteors. The radiant moves toward Castor (Alpha Geminorum) as maximum is approached. Reliable observations of the Geminids date back to 1862; since then there has been a rise in activity to its current high level. Minor planet (3200) Phaethon closely matches the Geminid stream, and it is likely to represent the dark-crusted burned-out nucleus of the once-active cometary progenitor of the Geminids.

Observing meteors

Meteor watching requires no equipment but a keen pair of eyes. Some showers produce fast, brilliant meteors, others produce slow-moving meteors; some meteors leave persistent glowing trails in their wake. Meteor colors are commonly reported, due to the way the eye perceives flashes of light and the actual hue produced by a meteor's burn-up (which depends on its speed and composition). Fireballs are

► *Brilliant fragmenting fireball, observed by the author on April 17, 1984, at 01:18 UT.*

exceptionally brilliant meteors brighter than magnitude −4 (as bright as the planet Venus at its brightest). They are produced by the atmospheric burn-up of larger meteoroids; as the outer surfaces of the meteoroids melt and ablate, streams of glowing particles are strewn along their wake.

To get a taste for meteor observation, a watch of an hour or so on the predicted maximum date of one of the more productive meteor showers will usually prove to be a satisfying experience. It will enable the observer to practice the techniques of recording meteors – making written notes, verbal notes (on a tape or digital voice recorder) and producing field drawings of the paths and appearance of particularly noteworthy meteors.

The generally observed activity of any shower is denoted by its zenithal hourly rate, which is the number of meteors an observer might perceive in an hour under dark skies with the radiant directly overhead. All things being equal, more meteors will be visible as the meteor's radiant climbs higher in the sky, so it is best to plan observations for when the radiant has risen at least 20° above the horizon.

TABLE 34 HOURLY RATE OF METEORS VISIBLE FOR DIFFERENT RADIANT ALTITUDES, GIVEN DARK TRANSPARENT CONDITIONS, FOR A ZHR OF 100.

Radiant altitude	Observed hourly rate
90°	100
70°	94
50°	77
40°	64
30°	50
20°	34
10°	17

Atmospheric conditions have a great bearing on the observed rate of meteors. Atmospheric haze, twilight, bright moonlight or light pollution will render fainter meteors invisible. It is recommended that meteor watches are conducted when the Sun is at least about 12° below the horizon. Moonlight presents a problem between first and last quarter phase, particularly if the Moon is high in the sky. Ten times more meteors are visible under dark moonless conditions than when observed under the light of a full Moon.

Meteor watching

Meteor watching – particularly on the dates predicted for shower maxima – can be an enjoyable activity in its own right, either observing on one's own or in a group. In terms of equipment, all that is required is an accurate timepiece (one that is easy to read under dark conditions), a dim red torch, pencils, a notepad and a portable tape recorder or digital voice recorder.

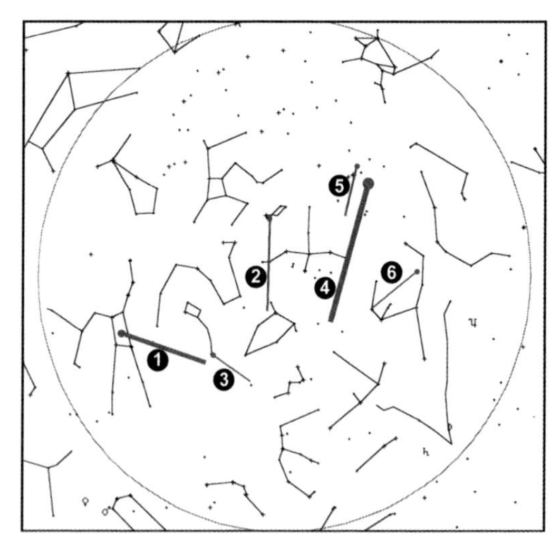

Perseid Watch 12 August 1998 21:10 - 22:20 UT Observer: Peter Grego, West Bromwich
Conditions: Direct light pollution on northeast horizon. Initially 30% cloud cover, clearing by 21:30, remained clear until 22:30 but complete cloud cover thereafter. Limiting magnitude: 4.5.

	Time	Magnitude	Speed	Colour	Train	
1	21:47:20	-1	Swift	Blue-white	None	
2	21:48:00	0	Swift	White	Slight	
3	21:49:40	1	Swift	White	None	
4	21:55:30	-4	Moderate	Blue-white	Lasted 1 sec	Brilliant meteor retained integrity and constant brilliance throughout flight
5	22:09:00	0	Very swift	White	None	
6	22:14:00	1	Swift	White	None	

◀ *Record by the author of a session of Perseid meteor observation on August 12, 1998, between 21:10 and 22:20 UT, with approximate plots of observed meteors against an all-sky star chart and their details below.*

A dark-sky site with a reasonably clear horizon is far preferable to a light-polluted urban location surrounded by buildings. If you can't make it out to a dark country site, a suitable corner of one's backyard will do. Comfort is essential: care should be taken to wrap up well in dry clothes, thermals, thick socks, sturdy boots and a hat – even in summer, night-time temperatures can become uncomfortably chilly. While much amateur astronomy is spent in awkward positions looking through telescope eyepieces at various angles, the meteor observer performs best when seated in a comfortable deck chair or on a reclining garden lounger with pillow support for the head; on winter nights a sleeping bag might not be such a bad idea (resist the temptation to drift off to sleep!). Though it may be a tempting prospect, it's inadvisable to observe in short sessions interspersed with visits indoors because the observer's eyes require at least 20 minutes to adapt sufficiently to the dark.

The observer's attention should be directed to an area of sky around 20° to 40° to one side of the radiant, and some 50° to 70° above the horizon, and it is best to follow this point on the celestial sphere throughout the observing session. The observing notes should specify the point in RA and Dec. Many observers tend to direct their attention straight toward the shower's radiant, but this actually reduces the number of meteors that might be observed. If, during the session, the observation point falls outside the parameters above, choose another to view, and note the new position. An attempt should be made to become familiar with the magnitudes of the stars within this general area, as this will help greatly in estimating the magnitude of observed meteors.

To secure a scientific record simply record the number of meteors viewed, noting the estimated magnitude and whether it belonged to the shower under observation or whether it was a sporadic. Notes can be made on a piece of paper, a tape recorder or a digital voice recorder.

Meteor observing notes ought to specify the following:

- Times (in UT) of the start and end of the meteor watch, and any breaks in observing.
- The limiting magnitude, and any changes to it, during the session. (The limiting magnitude is the faintest star near the zenith visible with the unaided eye.)
- Details of any haze or cloud.
- The center of the field of view under observation, in RA and Dec.
- For particularly bright or unusual meteors, note their time, color, duration, track, persistence of train.

14 · MOON

Earth's natural satellite

Measuring 3476 km in diameter – around a quarter the diameter of the Earth – the Moon is our only known natural satellite. According to the most popular current theory of the Moon's origin, around 4.6 billion years ago a Mars-sized planet smashed into the newly formed planet Earth. In what was a glancing blow, the impactor's heavy core coalesced with the core of the Earth, while vast quantities of the mixed mantles of the Earth and its cosmic assailant were lobbed into space; much of this material quickly accreted to form the Moon.

Today the Moon's crust is cold and thick, its interior having long ago cooled to a point where its material is too sluggish to flow. Volcanism was rife on the young Moon, but internally fueled activity on the Moon's surface had dwindled away by around 3 billion years ago. High rates of asteroidal bombardment early in the Moon's history gouged out vast, imposing impact basins, many of which were filled with lava to produce the dark lunar maria (the Moon's "seas") visible on the Moon's near-side. Volcanic activity produced squat dome-like volcanic hills known as domes, and lava flows from volcanic vents produced winding valleys known as sinuous rilles. Smaller impacts produced the thousands of craters dotting the Moon's surface; the most recently formed ones are surrounded by bright rays of ejected material. Crustal tension in and around the maria gave rise to faults and rift valleys known as rilles; some rilles are straight, others curve around the peripheries of the larger impact basins. As the maria settled, compression produced dorsa – low ridges that snake across many of the marial plains.

◀ *Mare Crisium dominates the northeastern cusp of the 3-day-old lunar crescent, imaged through Sir Patrick Moore's 15-inch Newtonian by Pete Lawrence on March 14, 2005.*

There's no appreciable lunar atmosphere; no rain ever quenches the dust-dry lunar soil, and no wind ever stirs it. Erosion on the lunar surface is produced mainly by meteoritic impact, which sandblasts solid rock and wears it away to dust over long periods of time.

▲ *A comparison by the author between the Moon's apparent diameter at apogee and perigee. The perigee Moon has an area around 30% greater than that of the apogee Moon.*

The Moon's orbit

The Moon orbits the Earth every 27 days 7 hours 43 minutes, a period known as its sidereal month. Its orbit is slightly elliptical, ranging from around 356,400 km at its nearest to the Earth (perigee) to 406,700 km when furthest away (apogee). This produces a variation in the Moon's apparent diameter, from around 33.5 arcminutes at perigee to around 29.5 arcminutes at apogee; the perigee Moon has an apparent area some 30% larger than that of the apogee Moon. Each sidereal month, the Moon turns exactly once on its axis, rotating so that it keeps the same hemisphere turned in the Earth's direction; this is known as synchronous rotation.

Libration

Although it is locked in a synchronous rotation, a phenomenon known as libration causes the Moon to appear to rock a little from side to side during the course of the month, allowing a total of 59% of the Moon to be observed over a period of time; the remaining 41% of the surface constitutes the true far-side, unobservable from the Earth. Libration results from the elliptical shape of the Moon's orbit, allowing the terrestrial observer to peer a little beyond the mean edges of the Moon when circumstances are favorable.

The Moon's libration zones lie around the edge of the lunar disk on either side of longitude 90°E and 90°W. Over time, all the features within these zones are eventually librated around onto the near-side and make an appearance near the lunar limb, although their visibility depends on the Moon's phase. Lunar observers take libration into account when planning observations because even if a mean far-side

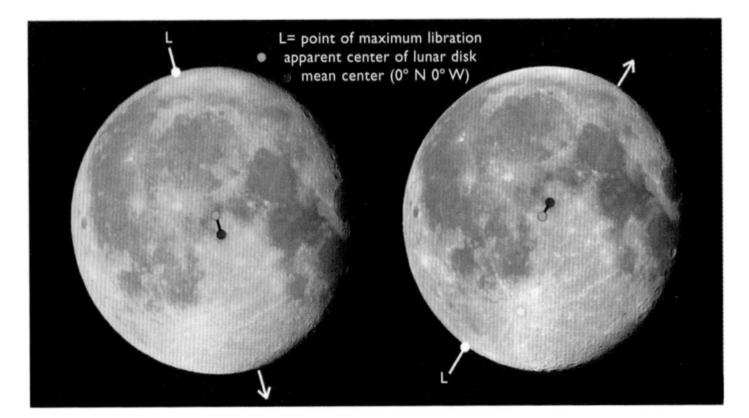

▲ *Lunar libration, illustrated by images taken by the author of the same phase separated by four lunations. In the image at left, which was taken on September 5, 2001, libration favors the northwestern limb, and the mean center of the Moon is displaced to the southeast. In the image at right, taken on December 31, 2001, libration favors the southwestern limb, and the Moon's mean center is displaced to the northeast.*

feature has been brought onto the Earth-facing hemisphere, foreshortening effects, combined with illumination circumstances, determine its practical observability. For example, there's little point attempting to observe the topographically interesting floor of the crater Einstein, a 170-km-diameter crater lying near the western limb, when there is a strong libration favoring the eastern lunar limb; even if libration favors the western limb, Einstein needs to be illuminated by a reasonably low Sun in order to show any topographic detail.

Libration often brings into view the north and south lunar polar regions. Under favorable conditions of libration and illumination it is possible to observe the actual surface position of the lunar north pole, which is located on the far northern rim of a small unnamed crater perched on the northern rim of the crater Peary. The Moon's south pole, however, lies on the floor of the crater Shackleton, obscured by intervening topography and unobservable from the Earth.

Libration does not greatly affect the apparent shapes of features nearer the center of the lunar disk, but it does affect when features appear on the Moon's terminator (the division between the day and night hemispheres of the Moon, marking the line of sunrise or sunset). For example, the large crater Ptolemaeus (2°W, 9°S) will have fully emerged into the sunlight of a first quarter Moon when there is a good libration to the east; when there is a strong libration to the west,

Ptolemaeus is completely immersed in shadow beyond the terminator at first quarter phase.

Numerous astronomical almanacs contain tables for the calculation of libration. Many good astronomical programs available for personal computers are capable of displaying libration and the lunar phase graphically, taking out the time-consuming mathematics involved in calculating and plotting an individual terminator.

Lunar phases

The Moon slowly moves from west to east, traveling an apparent angular distance a little more than its own diameter each hour. Gliding through the familiar constellations of the zodiac, in an orbital plane inclined some 5° to the ecliptic, the Moon travels just over 13° each day – about the width of your outstretched hand.

The Moon goes through a complete cycle of phases, from new Moon, through full Moon and back to new Moon in 29.5 days – a period known as its synodic month, or lunation. Since the Moon's orbital plane is near the ecliptic, its monthly path among the zodiacal constellations is similar to that followed by the Sun during a whole year. The first quarter Moon is 90° east of the Sun, located around where the Sun will be three months hence. The full Moon is on the opposite side of the sky to the Sun, in the same area of sky where the Sun will be six months later. The winter solstice Sun reaches its lowest point above the southern horizon, while the midwinter full Moon

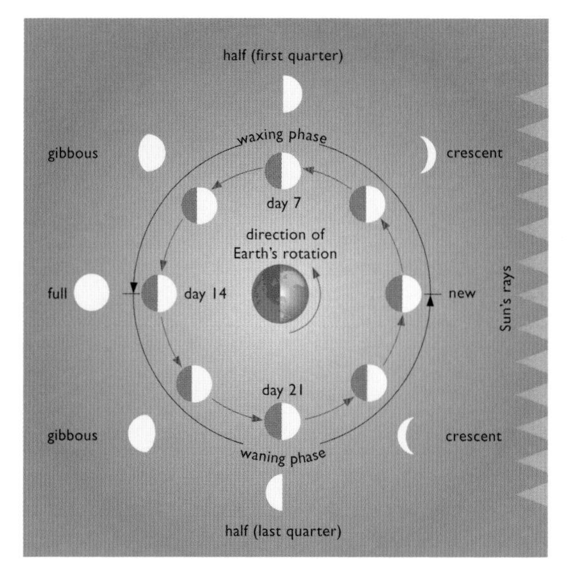

▶ *As the Moon orbits the Earth, its phase appears to grow, waxing from narrow crescent, through first quarter, gibbous to fully illuminated. The full Moon is opposite to the Sun. After full, the phase narrows, waning through gibbous, last quarter and narrow crescent. The new Moon lies in the Sun's direction and can only be seen during a solar eclipse.*

◀ *A clear view of features on the hemisphere of the Moon unilluminated by direct sunshine, but lit by faint earthshine. The bright lunar crescent has been overexposed to capture the phenomenon. This image was taken by Pete Lawrence on February 12, 2005.*

rides at its highest in the midnight skies. At summer solstice the Sun reaches its highest point above the horizon, while the midsummer full Moon appears at its lowest point above the horizon at culmination. The last quarter Moon is 90° west of the Sun, located around where the Sun was three months earlier.

Earthshine

When the Moon is a narrow waxing crescent in a reasonably dark evening sky, a faint illumination of the Moon's dark side can be discerned. Known as the earthshine, this phenomenon is caused by sunlight being reflected onto the Moon by the Earth. Viewed from the Moon's surface at this time, the Earth would appear as a bright waning gibbous sphere lighting the lunar landscape with up to 60 times the brilliance of a terrestrial full Moon. Using the unaided eye the earthshine may be followed for several days until around first quarter phase and picked up again in the last quarter of the lunation.

A general tour of the Moon's surface

Northeastern quadrant

Mare Serenitatis, a large circular marial plain some 600 km across, displays considerable well-defined tonal differences. The mare is bisected by a notably bright linear ray which passes through the crater **Bessel** (16 km); its origin appears to be the crater **Menelaus** (27 km), but it gives the visual impression of being a distant continuation of the rays from the prominent crater Tycho, some 2000 km away in the

TABLE 35 LUNAR NOMENCLATURE		
With the exception of craters, lunar features have two-part names, the first of which is a "descriptor" derived from Latin which identifies its type.		
Descriptor	**Meaning in Latin**	**Feature**
Catena	Chain	Chain of craters
Dorsum (pl. dorsa)	Back	Wrinkle ridge
Lacus	Lake	Small plain
Mare (pl. maria)	Sea	Large plain
Mons (pl. montes)	Mount	Mountain
Oceanus	Ocean	Very large plain
Palus	Swamp	Small plain
Planitia	Plain	Low plain
Promontorium	Promontory	Headland jutting into a mare
Rima (pl. rimae)	Crack	Rille (narrow valley)
Rupes	Cliff	Scarp
Sinus	Bay	Indentation at edge of mare
Vallis (pl. valles)	Valley	Large valley

Moon's southern uplands. Numerous large dorsa can be seen winding across the surface of Mare Serenitatis; formerly known as the "serpentine ridge," **Dorsa Smirnov** runs for some 130 km, joining with the 290-km-long **Dorsa Lister** in the south. Mare Serenitatis' walls are breached in the northeast by the irregular lava plain of **Lacus Somniorum** (Lake of Dreams, 340 km). Here can be found **Posidonius** (95 km), a beautiful crater with an interesting floor full of rilles and hills. Adjoining Lacus Somniorum in the north, **Lacus Mortis** (150 km) contains numerous impressive linear rilles, its center dominated by the crater **Bürg** (40 km).

Mare Serenitatis' southwestern border is marked by the parallel linear striations of the low-key mountain range **Montes Haemus**. The mare's western

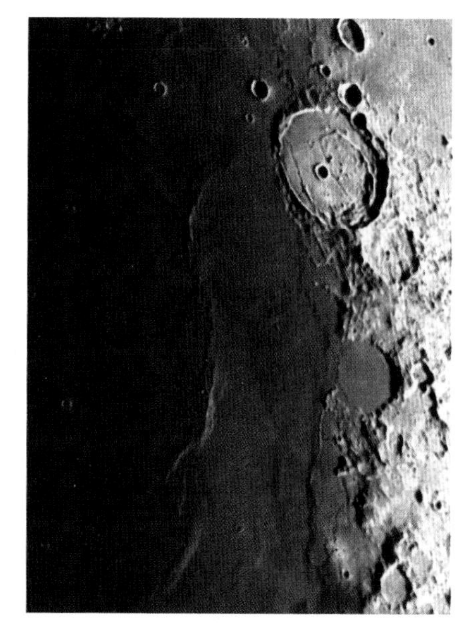

▶ *One of the Moon's most prominent wrinkle ridges, Dorsa Smirnov snakes across Mare Serenitatis. Image by Mike Brown, March 11, 2000.*

reaches connect with **Mare Imbrium** (Sea of Rains), and to the north can be found **Montes Caucasus**, a range which forms a mighty north–south wedge some 520 km long. Two large, impressive craters – **Aristoteles** (87 km) and **Eudoxus** (67 km) – stand proud in the highlands between Mare Serenitatis and **Mare Frigoris** (Sea of Cold). Beyond the far eastern end of Mare Frigoris is a similarly impressive duo, **Atlas** (87 km) and **Hercules** (69 km). To their northeast lies the dark, smooth-floored **Endymion** (125 km) and, nearer the limb, **Mare Humboldtianum** (Humboldt's Sea), a clearly defined mare whose visibility varies with the extent of libration.

Toward the eastern limb, a group of "stand-alone" maria include the large prominent oval of **Mare Crisium** (Sea of Crises, 620 × 570 km) and the smaller irregular-shaped patches of **Mare Anguis** (Serpent Sea, 130 × 30 km), **Mare Undarum** (Sea of Waves, 243 km), **Mare Spumans** (Foaming Sea, 139 km), **Mare Marginis** (Border Sea, 150 × 580 km) and **Mare Smythii** (Smyth's Sea, 580 × 550 km); the latter two straddle the 90°E line and their visibility is greatly affected by libration.

Mare Crisium is an imposing sight when it lies near the morning or evening terminator. In the morning, the large mountains towering above its western margin glint brightly in the Sun; in the evening, they cast broad shadows onto its floor. Immediately north of Mare Crisium can be found the large craters **Cleomedes** (126 km), **Burckhardt** (57 km), **Geminus** (86 km) and **Messala** (124 km). Just to the west of Mare Crisium, the small bright impact crater **Proclus** throws aside a remarkable fan-shaped array of bright ejecta material.

Approximately 700 km wide, the broad, square-shouldered **Mare Tranquillitatis** (Sea of Tranquillity) has the largest surface area of all maria in the northeastern quadrant. Its southwestern reaches – the site of the historic moonwalk of Neil Armstrong and Buzz Aldrin in 1969 – display a number of interesting features. **Lamont**, a sprawling mass of dorsa, can be seen only under conditions of low illumination when it is near the terminator. Arago Alpha and Beta, two very

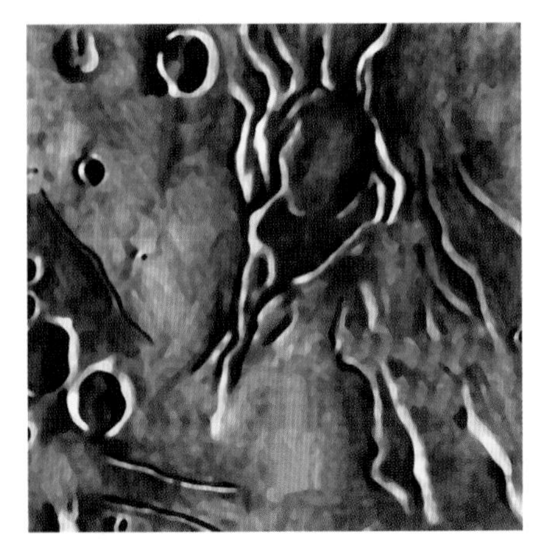

◀ *Lamont, an unusual feature composed of wrinkle ridges in Mare Tranquillitatis. Observation by Peter Grego.*

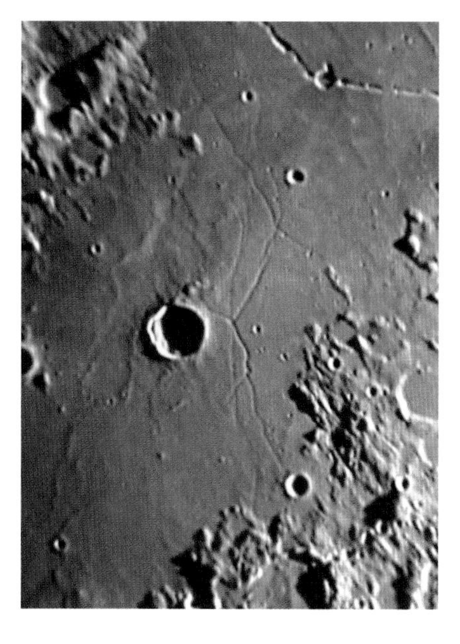

▶ *Intricate rilles around Triesnecker crater (near center) and the bold crater chain/rille Rima Hyginus (near top) can be viewed in northeastern Sinus Medii. Image by Mike Brown, April 16, 2005.*

large volcanic domes, lie near **Arago** (26 km), northwest of Lamont. Near the southwestern shore of Mare Tranquillitatis lie numerous linear and arcuate rilles. On the other side of the mare can be found a pair of smaller domes, **Cauchy Tau** and **Omega**, and the long curving fault lines of **Rupes** and **Rima Cauchy**.

Near the center of the lunar disk are the dark plains of **Mare Vaporum** (Sea of Vapors, 230 km) and **Sinus Medii** (Central Bay, 350 km). The topography in this area is fascinating to view, as much of it appears to be radially oriented to Mare Imbrium in the northwestern quadrant. Between Mare Vaporum and Sinus Medii, a fascinating collection of rilles can be studied telescopically. These include the intricate, interconnected mass of **Rimae Triesnecker** (covering an area some 200 km from north to south), the beautiful connected pearl necklace of **Rima Hyginus** (220 km), and the broad linear valley **Rima Ariadaeus** (220 km), which cuts across the highlands into the border of Mare Tranquillitatis.

Southeastern quadrant

Extending southeast of Mare Tranquillitatis, the diamond-shaped **Mare Fecunditatis** (Sea of Fertility, 600 km) is most notable for the double crater **Messier** (9 × 11 km) and **Messier A** (13 × 11 km) and its remarkable double linear rays which extend like narrow searchlight beams across the mare to the west. A number of wrinkle ridges (dorsa) adorn the mare, including several that appear to trace the outlines of ancient craters buried by lava flows. Beyond Mare Fecunditatis' eastern border, the large bright crater **Langrenus** (132 km) makes an impressive sight through binoculars and telescopes – during the first half of the lunation it can easily be discerned with the unaided eye as a bright spot. Langrenus, along with the large craters **Vendelinus** (147 km), **Petavius** (177 km) and **Furnerius** (125 km) to its south, make a prominent sight when the Moon is a young crescent or a waning gibbous phase. Petavius is worth examining closely; it has broad, magnificently detailed walls, an imposing central mountain massif and several large linear rilles on its floor.

On the southeastern limb, a large collection of lava-filled craters called **Mare Australe** (Southern Sea, 650 km) can be viewed under a favorable libration. Among the more noteworthy features near the southeastern limb is the huge disintegrated crater **Janssen** (190 km), its floor rifted with ancient rilles and overlapped by numerous younger craters. **Vallis Rheita** (500 km), one of the longest and broadest of all lunar valleys, is, like the narrower but equally lengthy **Vallis Snellius** nearby, a secondary impact feature radial to the Nectaris multiringed impact basin.

Mare Nectaris (Sea of Nectar, 350 km) is a small sea which flows from the southern reaches of Mare Tranquillitatis via the narrow straights of **Sinus Asperitatis** (Bay of Asperity). **Fracastorius** (124 km), a crater whose entire northern wall has been obliterated by mare lava flows, makes a prominent bay in southern Mare Nectaris. Immediately west of Mare Nectaris, a joined trio of large craters – **Theophilus** (100 km), **Cyrillus** (93 km) and **Catharina** (100 km) – demands attention when illuminated by the morning or evening Sun when the Moon is aged around 5 days and 19 days, respectively. Theophilus is particularly magnificent, with its broad terraced walls, sharp rim and huge cluster of central mountains. To the south, the giant curving scarp of **Rupes Altai** cuts through cratered mountainous terrain for around 500 km to the crater **Piccolomini**. Visible as a bright crenulated line under a morning illumination and casting a dark shadow eastward under an evening Sun, Rupes Altai marks the southwestern part of the outer ring of the large Nectaris multiringed impact basin.

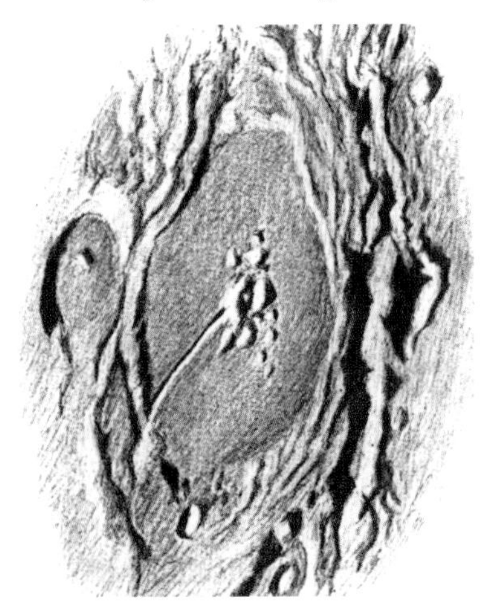

▲ *Petavius, with its large internal linear rille. Observation by Grahame Wheatley, October 17, 1997.*

Between Rupes Altai and the Moon's central meridian, a mass of craters, large and small, jostle to capture the observer's attention. Entirely devoid of dark marial patches, the lunar southern uplands are an enthralling sight through any instrument for most of the lunation.

It is easy for an observer to get lost in the area, even with a good lunar map. Large and impressive craters in the southeastern quadrant include **Albategnius** (136 km) with its smooth floor and prominent central peak, **Maurolycus** (114 km) with its superb double southern ramparts, and the collection of large craters including **Hommel** (125 km), **Vlacq** (89 km) and **Rosenberger** (96 km) near the southeastern limb.

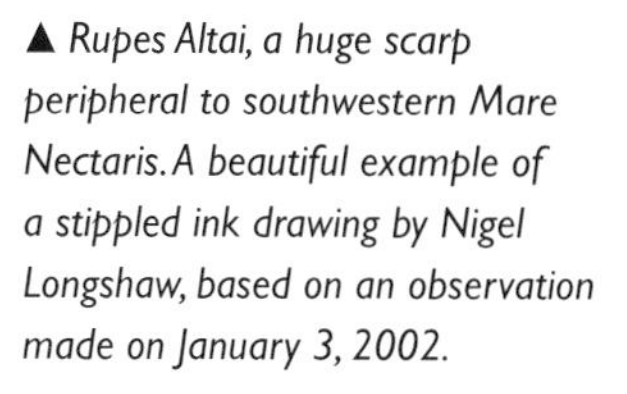

▲ *Rupes Altai, a huge scarp peripheral to southwestern Mare Nectaris. A beautiful example of a stippled ink drawing by Nigel Longshaw, based on an observation made on January 3, 2002.*

Northwestern quadrant

Mare Imbrium, the largest fully flooded marial basin on the lunar surface, dominates the northeastern quadrant of the Moon. Measuring some 1250 km in diameter, it is easily located with the unaided eye. The Imbrium plains are crossed by numerous dorsa, visible under a low illumination, and its southern half is pasted with the bright rays of crater **Copernicus**. Topographically, the most interesting part of Mare Imbrium is its eastern half, where a magnificent trio of craters can be viewed through binoculars – **Archimedes** (83 km), **Aristillus** (55 km) and **Autolycus** (39 km). With its smooth, flooded floor and low, rounded walls, Archimedes makes an interesting comparison with Aristillus and Autolycus, both of which have deep floors, central elevations, terraced walls and radial secondary impact structures. Nearby, the peaks of **Montes Spitzbergen** form a lovely compact mountain range poking out of the mare. South of Archimedes, a chaotic mountainous terrain covers a large area; the **Montes Archimedes** region can be discerned as a light patch using the unaided eye. The whole area is telescopically at its most magnificent when the Moon is around nine days old.

Mare Imbrium is surrounded for much of its circumference by a spectacular and near-continuous series of mountain ranges. Beginning in the northwest, **Mons Gruithuisen Gamma** and **Delta** form two massive, rounded mountain blocks – both are probably large volcanic domes. Hilly uplands to their north blend with the semicircular arc of

▲ *View of the large dark-floored crater Plato and mountain peaks jutting out from Mare Imbrium, including Montes Teneriffe and Mons Pico. Image by Mike Brown, September 21, 2000.*

Montes Jura, a range marking the northern border of **Sinus Iridum** (Rainbow Bay, 260 km in diameter), forming a deep bay on Mare Imbrium's northwestern shore.

Low, undulating highlands extend around the northern shore of Mare Imbrium, broadening to encapsulate the large dark-floored crater **Plato** (100 km). Jagged shadows cast onto Plato's smooth floor by its walls make fascinating viewing through a telescope as the Sun rises over the area. Several mountain ranges and isolated peaks rise out of Mare Imbrium, south of Plato, including the broad linear mass of **Montes Recti** (Straight Range, 90 km long), the scattered cluster of **Montes Teneriffe** (110 km long), **Mons Pico** and **Mons Piton**. Beyond the eastern end of Montes Alpes, the crater **Cassini** (57 km) has an unusual appearance, with a low, rounded outer flange beyond its sharp rim, and a flooded floor dented with two sizable bright craters.

East of Plato, the highlands grow into **Montes Alpes** (420 km long), whose peaks rise to average heights in excess of 2000 meters. The range is cleanly cut through by **Vallis Alpes**, a giant rift valley 130 km long and in places 20 km wide, whose steep ramparts rise some 2000 meters above its flooded floor. A small sinuous rille runs down the center of the valley floor, although at least a 150 mm telescope is required to see part of it. Rilles and other fault features cut neatly through parts of Mare Imbrium's mountain borders, and numerous rilles can be seen

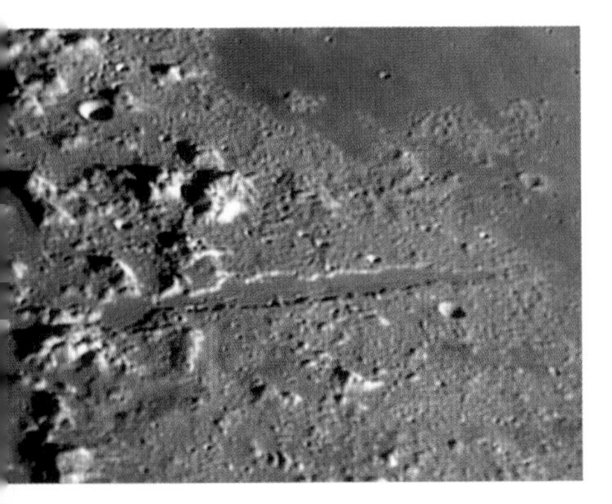

◀ *Vallis Alpes, a rift valley in the lunar Alps. Image by Jamie Cooper, March 19, 2005.*

within the sea itself, including **Rima Bradley** (130 km long), **Rima Fresnel** (90 km long) and the sinuous **Rima Hadley** (80 km long) near its southeastern shore, the site of the Apollo 15 landing.

▲ *The magnificent lunar Apennines, along with Hadley Rille nearby in Mare Imbrium. Image by Dave Tyler, May 27, 2004.*

Mare Imbrium's southeastern border is formed by the vast range of **Montes Apenninus**, some of whose peaks rise above 5000 meters. Much of the Apennines is scored and striated in a pattern radial to Mare Imbrium, best seen around first and last quarter phase. **Eratosthenes** (60 km) lies at the western end of the Apennines. After a gap in the mountain chain, **Montes Carpatus** marks Mare Imbrium's southern border. More than 400 km long, the range is overlain by ejecta from crater **Copernicus** (93 km) to its south.

Copernicus is one of the Moon's most magnificent impact craters, and our view of it from almost directly above certainly adds to its grandeur. Its inner walls are broad and terraced, and a group of mountains rises at the center of its floor. Copernicus' outer flanks are striated with radial ridges and furrows, among which can be discerned chains of secondary impact craters. Spreading for hundreds of kilometers away from Copernicus, its bright ray system can easily be discerned with the

► *Magnificent Copernicus, with its impressive dimensions, spectacular appearance and bright rays, dominates its part of the Moon. Image by Brian Jeffrey, March 1, 2004.*

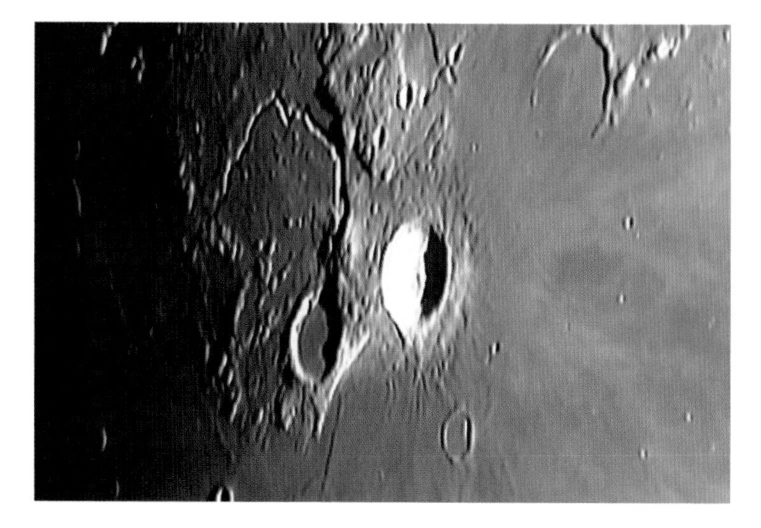

▲ *Brilliant Aristarchus and its fascinating environs emerge from the terminator as it sweeps over Oceanus Procellarum. The large curving valley to the west is Vallis Schröteri. Image by Anthony Ayiomamitis, October 25, 2004.*

unaided eye. West of Copernicus in the Hortensius–Milichius area, lie several groups of domes which are delightful to observe under a low illumination. Further west, across **Mare Insularum** (Sea of Isles), rays from the prominent crater **Kepler** (32 km) mesh with those of Copernicus.

Mare Imbrium blends with the gray plains of **Oceanus Procellarum** (Ocean of Storms), the largest lunar marial expanse, which covers a large portion of the near-side western hemisphere. Its most prominent resident, the crater Aristarchus (40 km) is one of the brightest features on the Moon. **Aristarchus**, with its brilliant terraced inner flanks and extensive ray system, makes an interesting comparison with nearby **Herodotus** (35 km), a dark, smooth-floored crater with low walls. To the north, the Aristarchus Plateau contains a variety of interesting topographic features, including **Vallis Schröteri** (160 km long), the biggest sinuous rille on the Moon, visible through small telescopes, the ghost crater **Prinz** (47 km) and **Rimae Prinz**, an assortment of smaller sinuous rilles, some of which are visible through a 150 mm telescope. Sprawling across the **Sinus Roris** (Bay of Dew) some distance north of the Aristarchus Plateau, the large and decidedly lumpy volcanic plateau **Mons Rümker** (70 km across) can be seen through small instruments when it has emerged from the morning terminator, though it soon fades into invisibility as the Sun rises.

South of the Aristarchus Plateau can be found the largest field of domes on the lunar surface. West of **Marius** (41 km), the dome field contains dozens of low, rounded hills and domes, in and around which wind numerous narrow sinuous rilles. **Reiner Gamma**, an interesting bright patch nearby, may mark the site of a recent cometary impact.

A number of large and topographically interesting craters lie near the limb of the northwestern quadrant. In the northwest are **Pythagoras** (130 km), with its prominent central peaks and intricate inner walls, and the ancient eroded craters **Babbage** (144 km), **South** (108 km) and **J Herschel** (156 km). Near the western limb lie the large craters **Hedin** (143 km) and **Hevelius** (106 km), both of whose floors are crossed by sizable linear rilles.

Southwestern quadrant

Oceanus Procellarum extends south of the lunar equator and flows into **Mare Nubium** (Sea of Clouds, 530 × 450 km) and **Mare Humorum** (Sea of Moisture, 360 × 380 km). A bay in eastern Mare Nubium is bisected by **Rupes Recta** (Straight Scarp), a normal fault some 110 km long, one of the largest of its kind on the Moon and certainly the most clean-cut. Once it has emerged from the morning terminator, it casts a broad black shadow onto the mare. Toward the end of the lunar day the scarp face catches the Sun and shines as a thin bright line. To its west, the sinuous rille **Rima Birt** (50 km long) makes a good test object for a 100 mm telescope. Southern Mare Nubium is host to the craters **Pitatus** (97 km), whose floor is bisected with rilles, and **Bullialdus** (61 km), with its magnificent radial impact structure. The flooded crater **Kies** (44 km) and its nearby summit-cratered dome are delightful to view telescopically.

East of Mare Imbrium, a trio of large craters running north–south near the Moon's central meridian grabs the observer's attention around first and last quarter phase. From the flooded crater **Ptolemaeus**

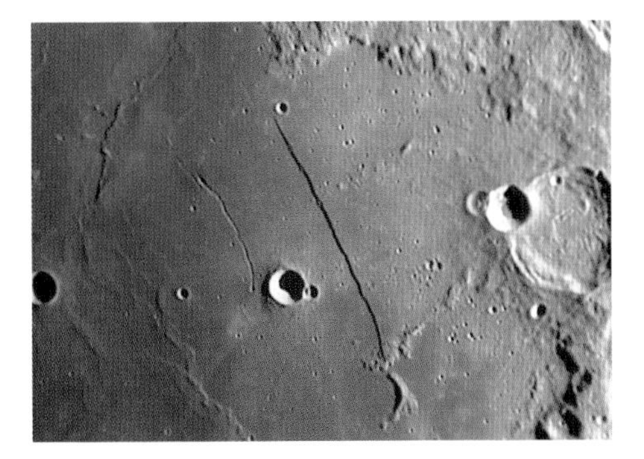

► *Rupes Recta, the biggest example of a textbook normal fault on the Moon. Image by Jamie Cooper, March 19, 2005.*

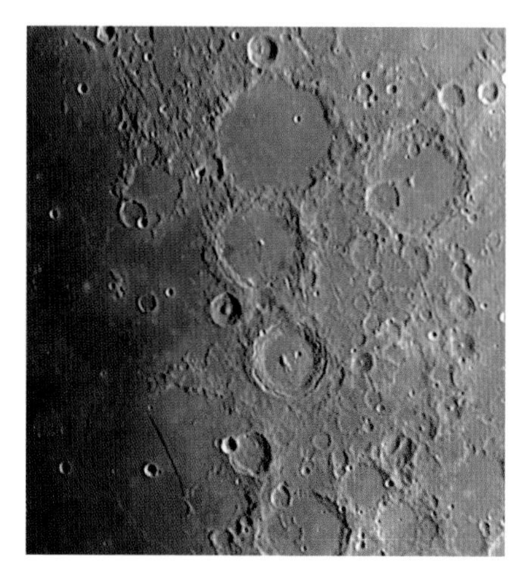

◀ *Ptolemaeus (large crater at top), Alphonsus (beneath it) and Rupes Recta (large dark line near bottom) make a splendid sight when the Moon is around 8 days old. Image by Jamie Cooper, March 19, 2005.*

(153 km) in the north, through adjoining **Alphonsus** (118 km) with its single central peak and fascinating cleft floor with dark-halo craters, to **Arzachel** (97 km) in the south, the scene is lunar grandeur at its best.

Rimae Hippalus, a parallel series of deep arcuate rilles, cut through the terrain east of Mare Humorum. Mare Humorum itself displays prominent concentric dorsa in the east, while its western floor and borders are rifted by **Rimae** and **Rupes Liebig** and **Rimae Mersenius**. The crater **Mersenius** (84 km) is remarkable for its smooth, apparently convex floor. One of the Moon's most oft-observed craters, **Gassendi** (110 km), lies on Mare Humorum's northern shore. It has a prominent central mountain complex surrounded by numerous linear rilles.

Clavius (225 km), nestled in the southern uplands, is highly impressive; an arc of five craters of decreasing size runs across its floor, from **Rutherfurd** (50 km) on Clavius' southern floor to **Clavius J** (10 km) in the west. Other large craters near the southern limb include **Blancanus** (105 km), the conjoined **Klaproth** (119 km) and **Casatus** (111 km), and **Moretus** (114 km). Some distance away, near the southern limb, lie the disintegrated impact basin **Bailly** (300 km) and the huge **Schiller–Zucchius** basin (325 km). Schiller itself is a fascinating elongated crater measuring 179 × 71 km. On the other side of the Schiller–Zucchius basin can be found the joined trio of **Phocylides** (114 km), a flooded crater; **Nasmyth** (77 km), whose southern section is overlapped by Phocylides; and **Wargentin** (84 km), a very unusual crater which is almost brim-full of lava. Wargentin adjoins the southwestern wall of **Schickard** (227 km), a crater notable for its relatively smooth, multitoned floor.

The very large multiringed **Orientale** impact basin straddles the 90°W line of longitude, and from time to time libration brings it into view at the Moon's southwestern limb. At its center, Mare Orientale (Eastern Sea, named before the 1961 International Astronomical

Union convention reversed east and west) is a circular patch of lava 300 km across. With a primary ring measuring around 930 km in diameter, the Orientale basin consists of several fractured mountain rings. Unlike the near-side basins, Orientale has been only partly flooded with lava. The dark lava tracts forming **Lacus Veris** (Lake of Spring) and **Lacus Autumni** (Lake of Autumn) are situated on the mean near-side and may sometimes be glimpsed even if Mare Orientale is not on view. Further north around the limb, the large dark-floored crater **Grimaldi** (222 km) lies centrally within a larger 430-km basin; to its west, **Riccioli** (146 km) has a partly flooded floor, cut through by the linear rilles of **Rimae Riccioli**.

Observing the Moon

Unaided eye views

Without optical aid the illuminated face of the Moon appears as a patchwork of dark and light areas. If the Moon is observed near full when it is close to the horizon, veiled by thin cloud or even seen through sunglasses, a surprising amount of detail can be discerned. One-third of the Moon's near-side is covered with maria, which appear as dark patches to the naked eye. The brighter areas surrounding them are mountainous cratered regions and the bright ray systems of a few large young impact craters.

It is possible to monitor some effects of libration with the unaided eye. The phenomenon is noticeable by observing the position of the marial strip of Mare Frigoris in the north and the presentation of Mare Crisium near the Moon's eastern limb. People with average eyesight will find that both features, when libration takes them close to the lunar limb, are something of a challenge to discern.

A list of lunar features to test visual acuity, compiled by the astronomer William H. Pickering last century, remains valid today. Starting with the easiest, and becoming progressively more difficult, the naked eye features are:

1. Bright region around Copernicus
2. Mare Nectaris
3. Mare Humorum
4. Bright region around Kepler
5. Gassendi region
6. Plinius region
7. Mare Vaporum
8. Lubiniezky region
9. Sinus Medii

10	Faintly shaded area near Sacrobosco
11	Dark spot at foot of Mons Huygens
12	Montes Riphaeus

Most people can discern Mare Nectaris with ease, and those with good eyesight will be able to make out Mare Vaporum with little difficulty. The Riphean Mountains are a really tough challenge even for those with excellent eyesight because the light gray cluster of peaks subtends an apparent angle of just 2 arcminutes. Irregularities along the Moon's terminator can sometimes be seen with the naked eye, notably the large crater Clavius around 8 days into the lunation or the bright curving arc of Montes Jura a day later, but better than average eyesight is required to see them.

Binocular observations

Viewed through even modest sized optical equipment, the Moon's surface resolves into a remarkable collection of seas, mountains and dozens of craters. A pair of small and unsophisticated opera glasses (which have the same basic optical configuration as Galileo's telescope) will magnify the Moon so that its real status as a rugged globe becomes apparent.

Many lunar observers remain quite happy to use a steadily held pair of binoculars in their pursuit of the Moon's wonders – and why not? If the binoculars are of sound optical quality – whether they are diminutive 8 × 20s or a giant pair of 15 × 80s – they will show enough detail to enable the observer to follow the appearance of the larger features as they emerge into the lunar sunshine. On the whole, binoculars have the advantage of being less expensive, easier to transport and to use, more able to withstand knocks and more useful for terrestrial purposes than telescopes. The Moon passes by some lovely star groupings during the lunar month, and binoculars are the best way of viewing the Moon in a low-power wide field. Moreover, it's best to use both eyes to gain the maximum aesthetic effect. The use of a simple steadying rod or a lightweight photographic tripod will increase the effectiveness of binoculars and enhance anyone's enjoyment of Moon watching.

Small telescopes

Examples of just about every type of lunar feature – faults, valleys, wrinkle ridges and domes included – can be seen through a telescope as small as 60 mm. As the morning terminator rolls back to uncover the lunar landscape, topographic features stand out boldly, their relief exaggerated because of the low angle of illumination. Small, rounded hills can appear to cast dramatically sharp shadows when near the terminator, and the shallowest of craters take on the appearance of deep pits. As the Sun climbs higher, the shadows recede, and the true vertical dimensions of the lunar landscape become apparent by midmorning. Under a high Sun, many lunar features that were prominent near the terminator can appear completely washed out, their outlines difficult to detect. Other features remain visible throughout the lunar afternoon, their edges brightly outlined or their dark floors clearly defined. A fortnight after they first emerged into the Sun, lunar features are immersed into the blackness of the evening terminator, and they endure two weeks of the cold lunar night. The sunrise and sunset days given for of the "Top 20 Lunar Attractions" in Table 36 are general guides only (libration offsets features considerably from the mean).

▲ *Map by the author showing the location of the 20 selected lunar features given in the table on page 204. 1. Petavius; 2. Vallis Rheita; 3. Rupes Altai; 4. Dorsa Smirnov; 5. Lamont; 6. Vallis Alpes; 7. Montes Apenninus; 8. Rima Hyginus; 9. Rimae Triesnecker; 10. Ptolemaeus; 11. Alphonsus; 12. Plato; 13. Montes Teneriffe & Mons Pico; 14. Tycho; 15. Rupes Recta; 16. Copernicus; 17. Gassendi; 18. Vallis Schröteri; 19. Aristarchus; 20. Mons Rümker.*

TABLE 36 TOP 20 LUNAR ATTRACTIONS				
Feature	Type	Dimensions	Sunrise	Sunset
Petavius	Impact crater	177 km diameter	3d	17d
Vallis Rheita	Secondary impact valley	500 km long, <30 km wide	4d	18d
Rupes Altai	Basin margin scarp	480 km long, <1000 m high	5d	19d
Dorsa Smirnov	Wrinkle ridge	130 km long	5d	20d
Lamont	Wrinkle ridge formation	75 km diameter	5d	20d
Vallis Alpes	Graben (rift valley)	130 km long, floor <1700 m beneath surrounding mountains	7d	21d
Montes Apenninus	Basin margin mountain range	1000 km long, individual peaks rising <5000 m	7d	21d
Rima Hyginus	Chain crater/rille	220 km long	7d	21d
Rimae Triesnecker	Network of rilles	200 km N–S, 50 km E–W, covers an area around 10,000 sq km	7d	21d
Ptolemaeus	Flooded crater	153 km diameter, floor sunk 2400 m beneath rim	7d	22d
Alphonsus	Crater	118 km diameter	7d	22d
Plato	Flooded crater	101 km diameter, floor 2000 m beneath crater rim	8d	22d
Montes Teneriffe/ Mons Pico	Mountain group	110 km long, peaks reach heights <2400 m	8d	22d
Tycho	Young ray crater	85 km diameter	8d	22d
Rupes Recta		110 km long, scarp face <300 m high, gradient 7°	8d	22d
Copernicus	Young ray crater	93 km diameter, 3760 m deep, central peaks <1200 m high	9d	23d
Gassendi	Flooded crater	110 km diameter, floor 1860 m beneath rim	10d	25d
Vallis Schröteri	Sinuous rille	160 km long, <10 km wide, <1000 m deep	11d	26d
Aristarchus	Young ray crater	40 km diameter, floor 3000 m beneath rim	11d	26d
Mons Rümker	Volcanic dome	70 km wide	12d	27d

*Note:

B – binoculars
t – Small telescope (smaller than 100 mm)
T – Large telescope (larger than 100 mm)
The sunrise and sunset days are general guides only; libration offsets features from the mean.

Location	Notes	Recommended minimum instrument*
25.3°S, 60.4°E	Large linear rille on floor	t
42°S, 51°E	Largest valley on near-side	t
24°S, 23°E	Largest scarp on near-side	t
25°N, 25°E	Prominent wrinkle ridge in Mare Serenitatis	t
5°N, 23.2°E	Unusual wrinkle ridge formation in Mare Tranquillitatis	t
49°N, 3°E	Largest graben on near-side	t
20°N, 3°W	Largest mountain range on Moon	B
7.8°N, 6.3°E	Chain of craters linked with a linear rille, central to which is Hyginus (11 km)	T
5°N, 5°E	Extensive interlinked rille system east of Triesnecker (26 km)	T
9.2°S, 1.8°W	Lava-flooded crater, saucer-like depressions visible on floor through large instrument	t
13.4°S, 2.8°W	Large, partly flooded crater with big central peak, floor rilles and dark-halo craters	t
51.6°N, 9.3°W	Large flooded crater with a rather smooth dark floor	t
48°N, 13°W	Group of mountains in Mare Imbrium	T
	Prominent ray crater in the Moon's southern uplands	t
22°S, 7°W	Clean-cut normal fault in Mare Nubium	t
9.7°N, 20°W	Imposing crater with terraced walls and central peaks	t
17.5°S, 39.9°W	Crater on northern border of Mare Humorum with large central peaks and a flooded floor full of linear rilles	T
26°N, 51°W	Moon's largest sinuous rille	T
23.7°N, 47.4°W	Brightest crater on near-side, has prominent rays and banded inner terraced walls	t
41°N, 58°W	Moon's biggest volcanic dome complex	t

Recording the Moon

Although the Moon's surface has now been mapped by spaceprobes in exquisite detail, there remain many good reasons for making observational drawings of lunar features. There is, of course, no significant Moon-mapping role left to the amateur astronomer, but there is still plenty of opportunity to discover previously unsuspected topographic features, particularly when features of very low relief are located close to the terminator and their presence is revealed by the shadows that they cast. Far from supplanting the visual observer, high-resolution CCD images provide an excellent starting point from which to pursue individual avenues for observational research.

Anyone making the effort to make accurate lunar observational drawings will discover an immensely useful and rewarding activity which improves every single aspect of his or her observing skills. The observer's ability to discern very fine detail constantly improves with the amount of time spent at the eyepiece. By making drawings of the Moon's features, observers learn to attend to detail instead of allowing their eye to wander onto the more obvious features. The discipline of accurate lunar drawing pays dividends in other fields of amateur astronomy requiring pencil drawing, like observing and recording features on Mars and Jupiter. After a while, the once-alien lunar landscape, with all its seemingly tongue-twisting nomenclature, becomes a known place; the apparent confusion of topography becomes increasingly familiar.

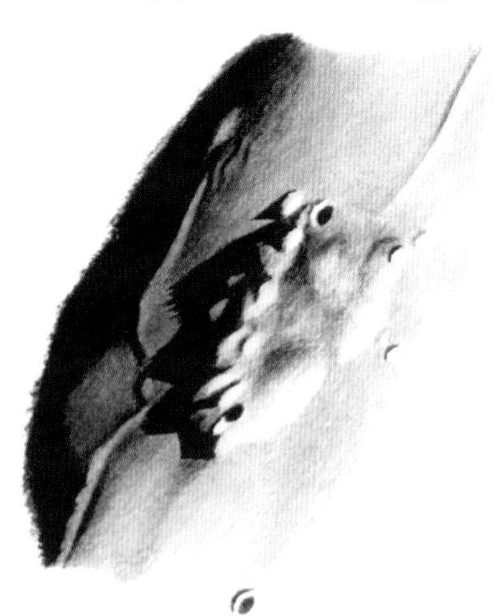

▲ *Mons Rümker, a large volcanic complex in Oceanus Procellarum. Observation by Grahame Wheatley, January 19, 2000.*

Pencil sketches

When making a drawing of the Moon, observational honesty and accuracy count above all. The enterprise ought to be seen as a scientific process, and not as a nocturnal art class – marks are not given for artistic flair.

With the Moon sharply focused in the eyepiece, the sheer wealth of detail on offer through even a small telescope can appear daunting. It is important to orientate yourself to begin with, and to discover which features are visible using a good map of the Moon. The eye is naturally drawn to the terminator, where most detail is visible due to the low angle of the sunlight illuminating that region. The area you choose to depict should ideally be quite small, such as an individual crater. If the chosen feature is not marked on the map then make a note of identifiable features nearby and indicate their positions in relation to the unknown area.

If possible, go back indoors and (copying from the map or a good photograph) make a reasonably large light outline drawing of your chosen area. This should be at least 75 mm across, larger if you are attempting to portray a region full of fine detail. Completing the first stage of a drawing indoors in this way saves a lot of time and allows the observer to concentrate upon filling in lunar detail at the eyepiece instead of worrying about depicting the exact configuration of the various features.

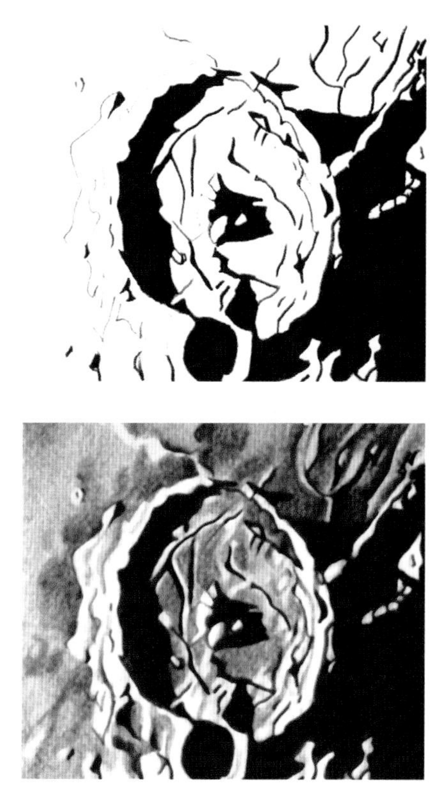

▲ *Observational drawing of Gassendi by the author, February 20, 2005.*

About an hour or two is the right amount of time to allocate per drawing session. Patience is vital because a rushed sketch is bound to be inaccurate, inevitably leading to frustration. At the eyepiece, basic outlines are drawn lightly at first, allowing corrections and erasures to be made if the need arises. Mark the outlines of the darkest, shadow-filled areas; when filling them in, apply minimal pressure and shade in layers, not in one frenzy of pencil pressure. General tonal shading is next applied, along with indications of fine detail, such as intricate terracing on crater walls, narrow rilles, small craters and so on. Subtle tonal shading is best left until last.

Unusual and interesting features should be indicated by making short written notes or by using a tape recorder or digital voice recorder to dictate observing notes while at the eyepiece. Each drawing should be accompanied by the usual observing information, such as date, start and finish times (UT), instrument and seeing. Other information, like the Sun's selenographic colongitude (a figure which indicates the height of the Sun above the lunar horizon at the time of observation), may be calculated later using a suitable ephemeris and written into the observing notes.

Line drawing

Features can be depicted in simple line form as an alternative to making shaded drawings. Bold lines represent the most prominent features like crater rims and the sharp outlines of black lunar shadows. More subtle features like low lunar domes and delicate detail are best recorded with light thin lines; dashed lines can show features like rays, and dotted lines can mark the boundaries of areas of different tone. Line drawings require plenty of descriptive notes, more so than a good toned pencil drawing. The method has the advantage of being quick and requires the minimum of drawing ability, and (used in conjunction with making intensity estimates, see below) it has the potential to be as accurate and as full of information as any toned pencil drawing.

Intensity estimates

During the lunation, the appearance of lunar features changes dramatically. Around a week after its shadow filled appearance at the terminator, a typical lunar crater has no shadow at all, and none of its relief detail is directly visible. Instead, actual brightness variations of the Moon's surface are seen. A line drawing of a crater made under these conditions is a useful complement to one made when the crater is full of shadow near the terminator.

The intensity estimation technique is to sketch the boundaries of areas of different brightness, and to make an estimate of those brightnesses on a scale of 0 to 10. The examples of tones described below are for a general low power telescopic view (×40). Each area will, of course, break down into further tonal variations under higher powered scrutiny.

0	Black – for the darkest of lunar shadows
1	Very dark grayish black – dark features under extremely shallow illumination
2	Dark gray – the southern half of Grimaldi's floor
3	Medium gray – the northern half of Grimaldi's floor
4	Yellow gray (subtle) – general tone of area west of Proclus
5	Pure light gray – general tone of Archimedes' floor
6	Light whitish gray – the ray system of Copernicus
7	Grayish white – the ray system of Kepler
8	Pure white – the southern floor of Copernicus
9	Glittering white – Tycho's rim
10	Brilliant white – the bright central peak of Aristarchus

Although these tints may be a little hard to picture, with practice they can be estimated with surprising consistency. The method is a little like that used by planetary observers, and although intensity estimates of this type depend heavily on the individual observer, excellent results can be obtained if the observations of a large number of people are combined.

Lunar programs for the PC

Many good planetarium-type computer programs provide ample information about the Moon, its illumination and phase, rising and setting times, and so on. A number of dedicated lunar programs are available which provide more detailed information to plan and research observations. They include: accurate ephemerides of lunar data, including data on libration in latitude and longitude, the Moon's phase and the Sun's selenographic colongitude; a graphic representation of the Moon showing all the major features, the position of the terminator and the effects of libration for any date and time input; a capability to display information and graphics on solar and lunar eclipses, and occultations of stars and planets.

Eclipses

If the Moon's orbit lay in precisely the same plane as the ecliptic (the path followed by the Sun through the sky), there would be a solar eclipse each new Moon and a lunar eclipse every full Moon. Because the plane of the Moon's orbit is inclined 5° to the ecliptic, the Moon usually passes a little to the north or south of the Sun at new Moon, and a little north or south of the Earth's shadow at full Moon. From time to time, the Moon does precisely align with the Sun and the

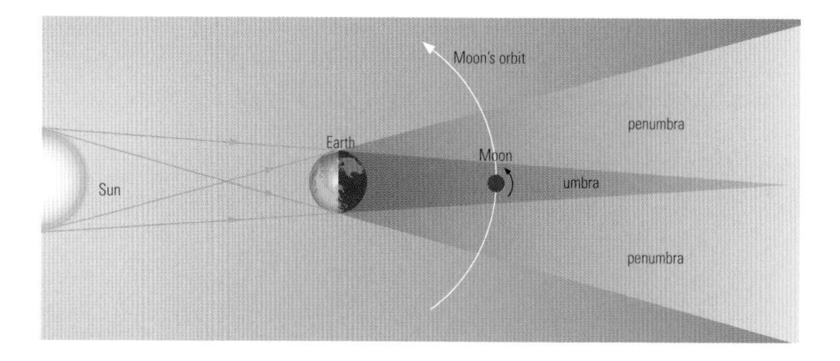

▲ Occasionally the Moon passes through the shadow cast into space by the Earth. This produces a lunar eclipse. Depending on how deeply the Moon enters the shadow, eclipses can be penumbral, partial or total.

Earth, producing both solar and lunar eclipses (see the next chapter for details of solar eclipses). Orbital geometry dictates that the Earth can experience a maximum of either five solar and two lunar eclipses, or four solar and three lunar eclipses within a single year.

There are two components to the Earth's shadow: a very dark umbral shadow cone, within which the Sun's direct light is completely obscured, and a fainter penumbral shadow, inside which only part of the Sun appears obscured. The Earth's umbral shadow cone is more than 1.3 million km long, and at the distance of the Moon a section through it is more than 9000 km across; the penumbra is around 17,000 km wide.

Lunar eclipses come in three varieties, depending on how deeply the Moon plunges through the Earth's shadow. Penumbral eclipses happen when the Moon passes through the penumbral shadow only; they are pretty unspectacular events, usually producing only a vague darkening of the Moon, most noticeable at the limb nearest the center of the shadow. Partial lunar eclipses occur when the Moon completely enters the Earth's penumbral shadow, and then into part of the darker umbral shadow. The edge of the umbral shadow is clearly defined, and appears as a dark segment at the Moon's edge.

A total lunar eclipse is one of astronomy's most spectacular sights. At totality the Moon is completely covered by the umbral shadow, but the Moon never totally vanishes because sunlight is refracted onto its surface through Earth's atmosphere. The maximum possible duration of a total lunar eclipse is 1 hour 42 minutes of totality. No two total lunar eclipses are the same; the hues, color distribution and intensity of the umbra always vary.

Steadily held binoculars deliver the best view of a lunar eclipse. Through binoculars the colors can be striking, and the sight of the Moon surrounded by a dark starry field gives a distinct three-dimensional impression. Telescopically, a field of view of at least one degree across is desirable; this will take in the entire Moon and allow nearby faint stars to be seen (and if necessary their occultations recorded) during totality. High-magnification eyepieces delivering a field of view that takes in only part of the eclipsed Moon are not necessary – at higher powers, the edge of the umbra loses its definition, making umbral contact timings with lunar features more difficult, and colors within the umbra less easy to discern.

Annotated line drawings on preprepared blanks may be attempted at 15-minute intervals. Notes should be made on the umbra's intensity and any colors within it, recording the width and definition of the edge of the umbra, along with any apparent irregularities in its outline. An estimate of the brightness of the totally eclipsed Moon can be made by comparing the apparent magnitude of the Moon viewed through the wrong end of binoculars with a star or planet of known magnitude.

Umbral shadow contact timings

By timing the passage of the edge of the umbra over certain prominent lunar features to the nearest minute, the size and shape of the Earth's shadow can be gauged. Recommended features to use in making umbral contact timings are mostly well-defined prominent bright spots, and include the craters Aristarchus, Kepler, Copernicus, Tycho, Plato, Manilius and Proclus. A contact timing should be made on the feature's immersion into the umbra and its emergence from it.

Photography

Pleasing results with simple photographic equipment can be obtained. An undriven wide-field multiple exposure showing the passage of the Moon through the sky and the progress of the eclipse can be obtained using a compact camera with a time exposure facility. With a driven telephoto lens a multiple exposure can be taken at various stages through the eclipse, including the Moon's immersion, totality and emergence from the umbra. Since the brightness of individual eclipses varies considerably, no hard guides can be given suggesting optimum exposures for totality. Using a conventional camera the total stages of a very dark

▲ *The lunar eclipse of October 28, 2004. This image, by Pete Lawrence, shows the immersion stages and the total phase of the eclipse.*

TABLE 37 PENUMBRAL, PARTIAL AND TOTAL LUNAR ECLIPSES 2006–2016

Date	Maximum (UT)	Type	Magnitude
2006 Mar 14	23:48	Penumbral	1.06
2006 Sep 07	18:51	Partial	0.19
2007 Mar 03	23:21	Total	1.24
2007 Aug 28	10:37	Total	1.48
2008 Feb 21	03:26	Total	1.111
2008 Aug 16	21:10	Partial	0.81
2009 Feb 09	14:38	Penumbral	−0.08
2009 Jul 07	09:39	Penumbral	−0.91
2009 Aug 06	00:39	Penumbral	−0.66
2009 Dec 31	19:23	Partial	0.08
2010 Jun 26	11:38	Partial	0.54
2010 Dec 21	08:17	Total	1.26
2011 Jun 15	20:13	Total	1.71
2011 Dec 10	14:32	Total	1.11
2012 Jun 04	11:03	Partial	0.38
2012 Nov 28	14:33	Penumbral	−0.18
2013 Apr 25	20:07	Partial	0.02
2013 May 25	04:10	Penumbral	−0.93
2013 Oct 18	23:50	Penumbral	−0.27
2014 Apr 15	07:46	Total	1.30
2014 Oct 08	10:55	Total	1.17
2015 Apr 04	12:00	Total	1.01
2015 Sep 28	02:47	Total	1.28
2016 Mar 23	11:47	Penumbral	−0.31
2016 Aug 18	09:42	Penumbral	−0.99

eclipse may require more than a minute's exposure, whereas a bright eclipse may register with just a few seconds. In any case, it is wise to bracket photographs and to experiment with exposures as the event unravels. The partial phases will need a little more exposure than if you were taking photographs of the Moon's regular phases to reveal anything of the tones within the umbra. Of course, on-the-spot experimentation with digital cameras used afocally at the telescope eyepiece is far easier than conventional photography, as the results can be viewed on the camera's electronic viewscreen immediately an exposure is taken.

An eclipse's maximum magnitude is reached when the Moon is nearest the center of the umbral shadow, and is measured in units of the Moon's diameter. Penumbral eclipses have a negative magnitude because the Moon does not enter the umbra. Partial eclipses have a magnitude between 0 and 1, while total eclipses range from 1 to a maximum of 1.888.

Transient Lunar Phenomena (TLP)

Through the centuries of telescopic lunar observation, transient phenomena have occasionally been observed on the Moon. Known as TLP (transient lunar phenomena), these events have taken the form of localized flashes, glows, obscurations, darkenings and changes of color. Unfortunately, TLP appear to occur on a temporary and unpredictable basis, although a number of specific features have been notably prone to TLP – perhaps it is no coincidence that many of these areas were once the site of volcanic or crustal activity along faulted mare borders. The region in and around the crater Aristarchus is by far the most TLP-prone area: more have been reported here than all the other features put together. These events have been chiefly brightenings, colored anomalies and obscurations of surface detail.

TABLE 38 FEATURES WITH REPORTED TLP ASSOCIATIONS			
Agrippa	Archimedes	Aristarchus	Atlas
Alphonsus	Bullialdus	Censorinus	Copernicus
Eratosthenes	Gassendi	Grimaldi	Herodotus
Kepler	Linné	Manilius	Mare Crisium
Menelaus	Mons Piton	Mons Pico	Picard
Plato	Posidonius	Proclus	Prom Laplace
Schickard	Theophilus	Tycho	Vallis Schröteri

Colored filters can enhance the visibility of anomalous colored areas on the lunar surface. By rapidly alternating the view through a red and a blue filter, faintly colored areas on the Moon tend to stand out more by appearing to flicker. A red area will appear brighter when viewed through a red filter, and darker when seen through the blue filter. The best results are obtained using a Wratten 25 red and a Wratten 38a blue filter; these can be mounted next to each other on a card and manually alternated in front of an eyepiece with a good level of eye relief. Areas displaying true coloration appear to flicker, while uncolored features of a similar albedo in the nearby region do not. A number of features appear to blink naturally, among them the southwestern part of the crater Fracastorius and a section of the western wall of Plato.

Solar hub

Around 5 billion years ago, a gravitational ripple within a cloud of cold interstellar gas and dust began the process of the formation of the Solar System. One of the denser parts of the cloud formed a dark nebula; collapsing inward under its own gravity, the nebula began to spin, producing a disk of material. At its center, the increasing pressure and temperature of the material finally grew so hot that thermonuclear reactions were triggered, and the Sun was born. Streaming away from the energetic young Sun, the solar wind blew away most of the dust and gas in the original nebula, leaving behind just the larger, denser clumps of matter which had formed mainly in the disk; these larger aggregations of material were to become the Sun's planets and their moons, asteroids and comets.

At a distance of around 26,000 light years from the center of the Galaxy, the Sun has a Galactic orbital period of 226 million years; it has made around 20 orbits around the center of the Milky Way since its birth. The Sun's orbital velocity around the Galaxy is some 217 km/s; it takes 8 days to travel 1 AU and 1400 years to travel 1 light year.

Dominating the hub of the Solar System, the Sun is an average-sized main-sequence star of G2 spectral class. Deep inside it, nuclear fusion converts hydrogen into helium, and energy is produced in the process – a staggering four million tons of matter is converted to energy each second. From the Earth the Sun has an apparent magnitude of −26.7 – a million times too bright to view without proper eye protection. The Sun's absolute magnitude is 4.8 (absolute magnitude is a star's brightness when viewed from a distance of 10 parsecs, or 32.6 light years); this is about the same as the absolute magnitude of Alpha Centauri, a nearby Sun-like star.

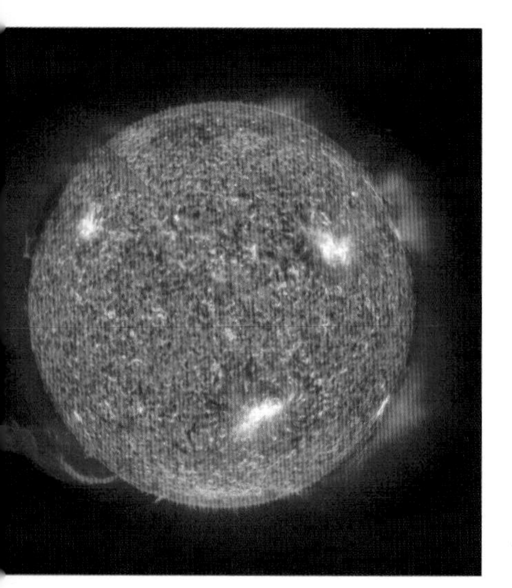

◀ *Our nearest star, the Sun, appears in all its magnificence when viewed in hydrogen-alpha (H-alpha) light – its red glowing chromosphere and limb prominences can easily be seen.*

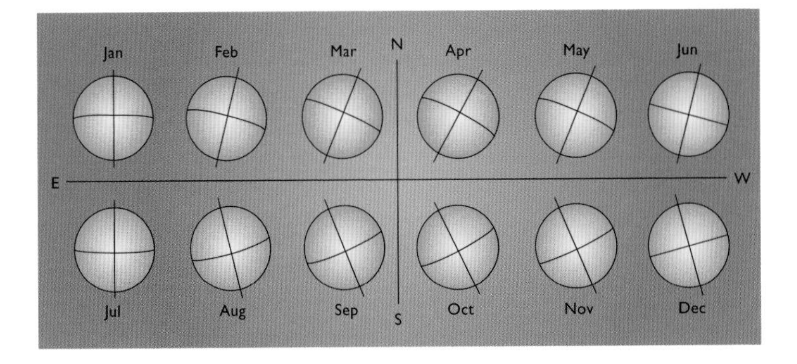

▲ *The apparent tilt of the Sun with respect to celestial coordinates throughout the year. The Sun itself rotates from east to west.*

Size, shape, axis and rotation

Measured across the photosphere (the bright surface of the Sun visible in ordinary light) the Sun is 1,392,000 km across at the equator. Being a ball of gas, the Sun's rotation period is differential (a phenomenon displayed by all the gas giant planets), varying between 25.6 days at the equator to 30.9 days at a latitude of 60° N or S (and even slower toward the poles). Such a slow rotation period means that centrifugal effects are relatively small. In fact, the Sun is almost perfectly spherical, its polar diameter being just 10 km shorter than its equatorial diameter.

The Sun rotates on its axis in the same direction as the planets orbit the Sun – anticlockwise, as viewed from above the Sun's north pole; in observational terms, features such as sunspots first appear at the Sun's eastern limb, transit the Sun's meridian more than 6 days later and disappear beyond the western limb after a further 6 days or so. The solar axis is inclined by 82.75° to the plane of the ecliptic (the plane of the Earth's orbit around the Sun), so the Sun's apparent tilt varies during the course of a year. The solar north pole is at is maximum tilt of 7.25° toward the Earth in September; six months later, in March, the solar south pole is at its maximum tilt of 7.25° to the Earth.

The Sun's ecliptic path

From our vantage point on planet Earth, the Sun appears to travel in an easterly direction along the ecliptic by a little more than one degree each day, completing one complete revolution around the celestial sphere every 365.26 days. The ecliptic passes through 13 constellations – all the ancient zodiacal constellations, plus Ophiuchus between Sagittarius and Scorpius. However, the Sun spends unequal amounts of time passing through each constellation along the ecliptic, since the

TABLE 39 CONSTELLATIONS ALONG THE ECLIPTIC			
Constellation	Extent along ecliptic	Entry	Duration of Sun's visit (days)
Aries	24.6°	Apr 18	25
Taurus	36.8°	May 14	37
Gemini	27.7°	Jun 21	28
Summer solstice		Jun 22	
Cancer	20.1°	Jul 20	20
Leo	35.7°	Aug 10	36
Virgo	44.1°	Sep 16	45
Autumn equinox		Sep 22	
Libra	23.1°	Oct 31	23
Scorpius	6.4°	Nov 23	6
Ophiuchus	18.9°	Nov 29	19
Sagittarius	33.4°	Dec 17	34
Winter solstice		Dec 21	
Capricornus	28.1°	Jan 19	29
Aquarius	23.8°	Feb 16	24
Pisces	37.2°	Mar 12	38
Vernal equinox		Mar 20	
*Cetus		Mar 27	

**Note: Since the ecliptic passes just 9 arcminutes north of the corner of the constellation of Cetus, the southern part of the Sun passes through Cetus for a period of around 13.5 hours from March 27 to 28 each year.*

constellation boundaries, officially fixed by the International Astronomical Union, do not coincide with those recognized in ancient times. The amount of space that each constellation occupies along the ecliptic varies from as little as 6.4° (Scorpius) to as much as 44.1° (Virgo), the Sun spending 6 to 45 days, respectively, traversing each of these constellations.

Seasons

Since the Earth's axial tilt is inclined by 66.5° to the plane of its orbit around the Sun, the ecliptic is angled at 23.5° to the celestial equator. As the Earth's axis remains fixed in space as it orbits the Sun, there is a cyclic variation in the apparent height of the Sun above the southern horizon at midday and the amount of daylight experienced at various places around the globe. The Sun's midday height above the horizon is at its highest in summer and at its lowest in winter. Regions bounded by the Arctic and Antarctic Circles lose daylight altogether during the winter, with the total period of winter darkness varying with proximity to the pole; at the poles themselves, there are six months of daylight and six months of night at around the winter solstice.

Winter solstice in the northern hemisphere (summer solstice in the southern hemisphere) takes place around December 21, the first day of the winter season, with the Sun in Sagittarius. Northern temperate latitudes experience their shortest period of daylight at winter solstice; from the UK this varies between around 8 hours in southern England, with the Sun 16° above the southern horizon at midday, to around 6.5 hours in northern Scotland, the Sun managing to reach a maximum altitude of just 9° above the southern horizon. Summer solstice in the northern hemisphere (winter solstice in the southern hemisphere) occurs around June 22, with the Sun in Gemini. Places in northern temperate latitudes experience their longest periods of daylight, while places in southern temperate latitudes have their shortest days. From northern Scotland there are some 18 hours of daylight, the Sun climbing to a maximum altitude of around 56° above the horizon; southern England experiences around 16.5 hours of daylight, with the Sun transiting the meridian at an altitude of around 63°.

In between the solstices are the vernal and autumnal equinoxes, the two points where the ecliptic and celestial equator intersect. Moving north, the Sun crosses the celestial equator at the vernal equinox around March 20; six months later, around September 22, the Sun crosses the celestial equator, moving southward, at the autumnal equinox. Spring begins on the day of the vernal equinox, while the autumn equinox marks the beginning of the autumn season. At the equinoxes, all parts of the globe have 12 hours of daylight and 12 hours of night (the word equinox means equal nights).

Sun–Earth distance

On average, the Sun is 149.6 million km from the Earth, around 400 times further than the Moon. This figure, known as an Astronomical Unit (AU), is the standard measurement astronomers use for all distances between objects in the Solar System (and, for that matter, distances between objects in other star systems). For example, Pluto is more than 50 AU (fifty times the average Earth–Sun distance) from the Sun at the furthest point in its orbit.

Like all the major planets, the Earth's orbit around the Sun is slightly elliptical, and Earth's distance from the Sun varies throughout the year. The Earth is at perihelion, 147.1 million km from the Sun, around January 2; aphelion takes place around July 2, when the Earth is at a distance of 152.1 million km. The Sun's apparent diameter varies between a maximum of 32′ 31″ when the Earth is at perihelion to a minimum of 31′ 27″ when at aphelion. The variation in the apparent diameter of the Sun becomes obvious during the phenomenon of a total solar eclipse; the type of eclipse observed (whether an annular or total eclipse) and the duration of totality is largely dependent on the apparent diameters of the Sun and the Moon. Long-duration totalities occur when the Sun's apparent diameter is on the small side and the apparent diameter of the Moon is as large as possible (see the section on solar eclipses below).

Solar activity

Activity beneath the Sun's 500-km-deep photosphere gives rise to a variety of observable phenomena, but it is not possible to view this activity directly. Everything above the photosphere is known collectively as the solar atmosphere, and is observable with a variety of telescopes tuned to wavelengths across the electromagnetic spectrum, from radio waves, through visible light to X-rays and gamma rays.

Most amateur solar observing centers around observations of sunspots in the photosphere made at visible light wavelengths, since

▼ Thermonuclear fission taking place within the Sun's intensely hot, pressurized core provides the source of its tremendous output of energy. Transmitted through the Sun's radiative and convective zones, sunlight finally bursts out at the photosphere, the visible outer layer of the Sun.

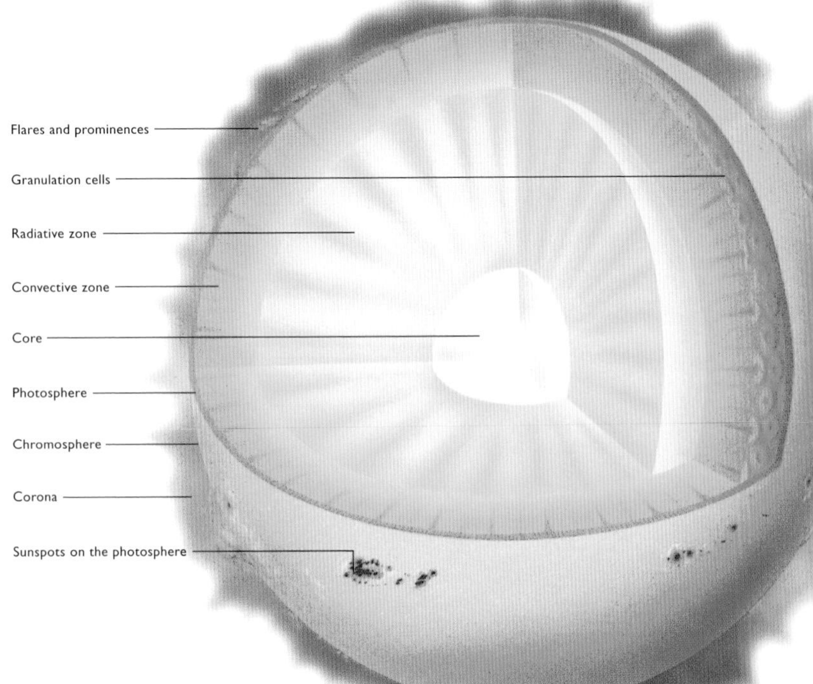

this requires little specialist equipment (though it does require special attention to safety – see below for advice on safe solar observing). Observations made in visible light may also reveal photospheric granulation, along with bright areas near the limb, called faculae. Observers with more sophisticated equipment tuned to hydrogen-alpha (H-alpha) wavelengths can observe the Sun's chromosphere – a thin transitional layer between the photosphere and the extensive outer atmosphere, the corona. In H-alpha, a host of phenomena can be viewed, including sunspots, granulation, plages, faculae, filaments, prominences and flares. A few observers pursue spectroscopic and spectrohelioscopic studies.

White-light features

Granulation

Huge convection cells of hot gases in the photosphere, each averaging around 1000 km in diameter, jostle for position in the photosphere as they rise and sink above the Sun's visible surface. These vast, hot bubbles produce a mottling effect called granulation. In white light, granulation is best viewed under good solar seeing conditions through telescopes larger than 100 mm. Due to the complexity of granulation and its transient nature, the visual observer cannot hope to portray it adequately on an observational drawing, though it may of course be noted; capturing granulation successfully on film or CCD can present something of a challenge.

Sunspots

Disturbances within the powerful magnetic field generated within the Sun cause sunspots to appear in the photosphere. The difference in temperature and brightness between the 5500°C photosphere and a sunspot's 3500°C interior makes the sunspot appear dark. If a sunspot were somehow removed from the solar surface, it would appear as a dazzlingly brilliant object. Sunspots are the most obvious signs of solar activity visible in normal "white" light.

Sunspots usually develop from small dark pores in the photosphere, and can take several days to grow to maximum size – on average, around 10,000 km in diameter. Fully formed sunspots often have two components – a darker central area called the umbra and a dusky peripheral region called the penumbra. Under good conditions, the penumbra may appear striated radially to the umbra – an effect caused by tubes of hot gas streaming into the spot from the surrounding photosphere. Sunspot behavior can be rather unpredictable. Some spots decay rapidly after a day or two on the solar disk, returning to their

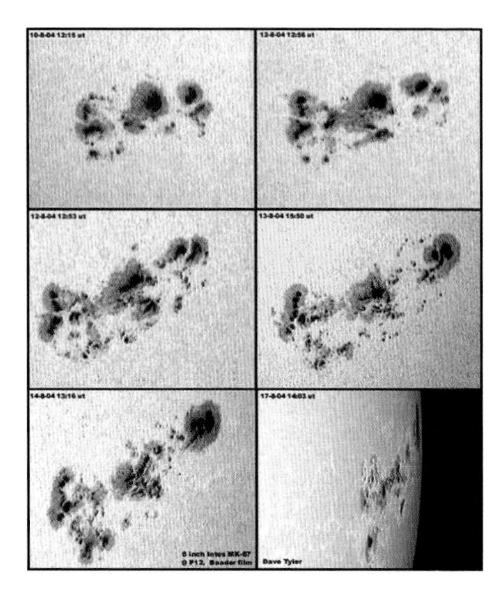

▶ *The development of a major sunspot group between August 10 and 17, 2004. The image sequence was taken by Dave Tyler using a 150 mm MCT (Intes MK-67) at f/12. A Baader whole-aperture solar filter was used.*

pore state and then disappearing. Other sunspots grow to enormous proportions – as large as 150,000 km across. They often appear in pairs (where the magnetic field lines loop above the photosphere and re-enter it many thousands of kilometers further to the east) or are attended by smaller groups of spots, with one large leader spot preceding the sunspot group. Large sunspots may last for many weeks, long enough for one or more rotations of the Sun.

Sunspots and the solar cycle

Records of sunspot numbers going back hundreds of years show that the Sun experiences an approximate 11-year cycle of activity, with individual cycles ranging from 7 to 17 years' long. Periods of high solar activity at solar maxima are characterized by an increased number of prominences and flares, leading to a greater frequency of geomagnetic storms and aurorae. Additionally, the solar cycle produces variations in the shape and intensity of the Sun's corona; at solar maximum, coronal streamers extend in all directions, while at solar minimum the corona extends mainly from the Sun's equatorial regions.

Detailed sunspot observations gathered since the late 19th century, including data on sunspot sizes and positions in addition to sunspot counts, demonstrate that sunspots do not crop up randomly over the Sun's surface. Instead, they develop mainly in two latitude bands on either side of the equator. A graph plotting sunspot positions for each solar rotation (known as the "Butterfly Diagram") clearly shows that these sunspot bands form at midlatitudes in the northern and southern hemispheres at the beginning of each cycle, and they widen and migrate toward the equator as the cycle progresses. Sunspots rarely appear at latitudes closer than 5° to the equator or nearer to the poles than 40° N or S. At solar maximum, sunspots appear nearer the equator and become more numerous – as many as 100 sunspots may be seen on the solar disk at one time. Many days can go by during solar minimum without any sunspots being seen.

Limb darkening

Images of the Sun taken in white light clearly show a phenomenon known as limb darkening, where the brightness of the Sun appears to diminish toward the edge of the solar disk. The effect is noticeable when viewing a projected image of the Sun, though it may not be obvious when the Sun is viewed directly through a telescope using a safe whole-aperture solar filter. Limb darkening is produced because the Sun's temperature and the amount of light it emits increases with depth. On looking toward the center of the solar disk, an observer is viewing the deepest and hottest layers within the Sun – those that emit the most light. Toward the Sun's edge, however, only the upper,

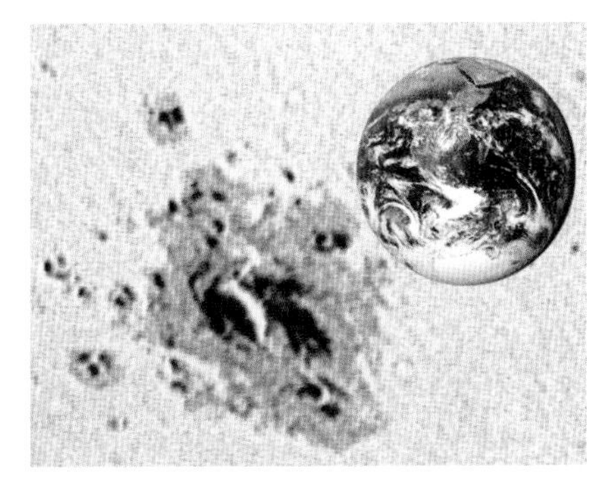

► *Sunspots are enormous objects, many of them larger than the diameter of the Earth, as shown in this comparison. The sunspot image was taken by Richard Bailey (NOAA 484, October 27, 2003, at 13:32 UT).*

▼ *A butterfly diagram drawn by the author showing the intensity and latitude of observed solar activity from 1870 to 2005.*

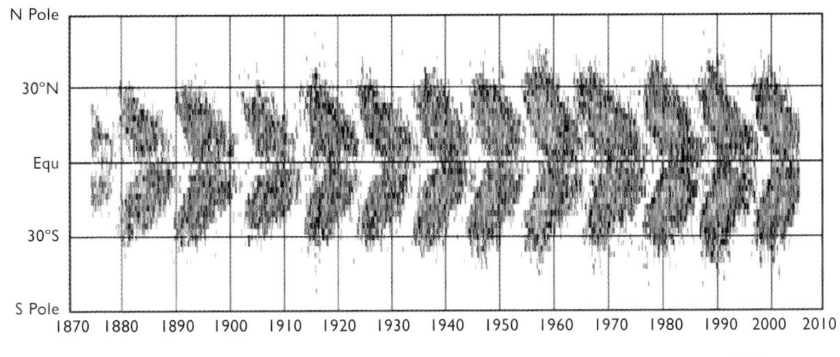

cooler layers – those that produce less light – are visible. The presence of limb darkening is proof that the Sun is not a solid sphere of burning material.

Faculae

Also associated with magnetic effects, faculae (meaning "little torches") are bright areas seen near the Sun's edge, often running in lengthy sinuous bands; limb darkening makes them easier to see in white light. Their origin is not well-understood.

Hydrogen-alpha features

Chromosphere

Around 8000 km thick, the chromosphere is a thin layer between the photosphere and the corona. Its temperature ranges from 6000°C to 20,000°C. Features within the chromosphere and beyond are visible in H-alpha wavelengths.

Prominences and filaments

The ability to discern prominences around the Sun's limb is one of the main reasons why many amateur astronomers invest in expensive H-alpha telescopes. Prominences are glowing clouds of hydrogen gas extending from the chromosphere into the lower corona to distances of up to 300,000 km or more. Eruptive prominences, emerging from active areas on the Sun, can develop into fantastically complex streamers, loops, arches or sprays, which may extend many arcminutes away

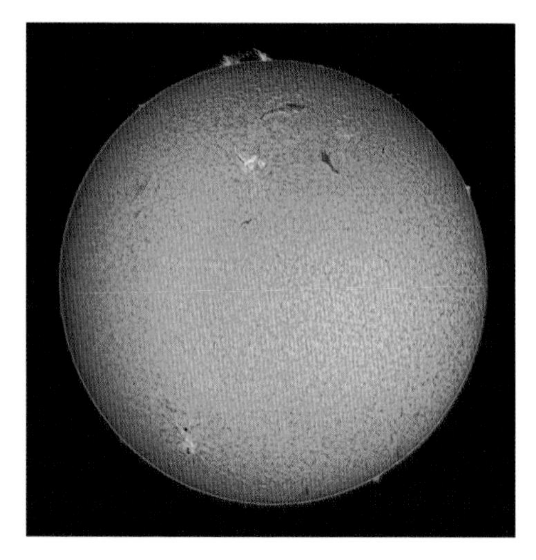

◀ *The Sun's disk, as imaged in H-alpha light on June 20, 2005, by Sean Walker.*

from the Sun's limb. Their development can be quite startling to view; over a period of a few hours they burst away from the Sun's limb, traveling at a velocity of 1300 km/s. The most violent events produce coronal mass ejections, which can produce a significant impact on the Earth's geomagnetic field, causing aurorae and communications disruptions.

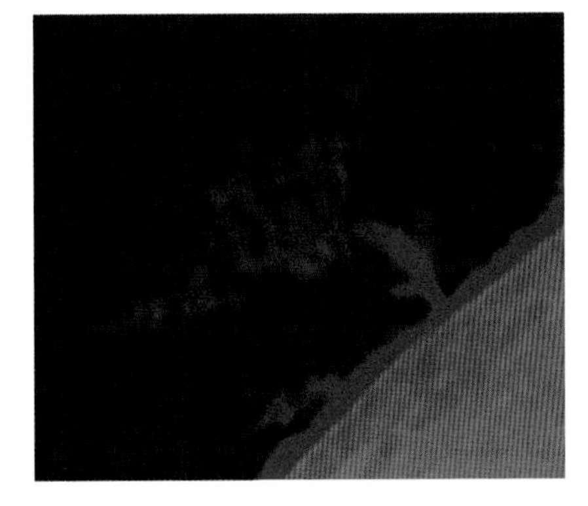

▲ *Solar prominence on the northeastern limb, imaged in H-alpha light by Richard Bailey on April 28, 2004, at 13:14 UT.*

Quiescent prominences are more stable features, often taking the appearance of hedgerows, curtains, floating arches and fans; they hang suspended in the corona, where they slowly develop, becoming more elongated during the course of a few solar rotations.

When they appear against the solar disk, both types of prominence look like dusky filaments snaking across the face of the Sun.

Associated with sunspot activity, prominences are clouds of partially ionized hydrogen gas in or slightly above the Sun's chromosphere; they are contorted into shapes, held in position and anchored to the Sun's surface by local magnetic fields. When the prominence's anchor points are disturbed by newly emerging magnetic fields of the same polarity, or by the decay of its own local field, its ties to the surface are loosened and it soars above the Sun's surface, lofted high by its magnetic buoyancy.

Spicules

Spicules are jets of gas around 1000 km across and 10,000 km long of just a few minutes' duration. They are most often seen on the limb in H-alpha light, and their appearance has been likened to a field of burning grass. Seen against the Sun's disk, they appear as dark spikes, usually far from any active areas. Spicules stem from the boundaries of granulation cells in the photosphere.

Plages

Plages are bright areas surrounding active regions (usually around sunspots); they are irregular in shape and vary in brightness. Plages mark emerging or reconnecting magnetic field lines, and they can last for several days.

Solar flares

Sudden eruptions known as solar flares are the most violent events on the Sun's surface. They usually last from a few minutes to several hours, and in that time they release unimaginable amounts of energy, equivalent to many millions of hydrogen bombs. Solar flares are more frequent near sunspot maximum, when smaller flares may appear on a daily basis and larger flares may take place with an average frequency of once a week. It is possible to observe the brightest flares in ordinary white light, but this is rather rare. Flares can give rise to shock waves moving at velocities of around 1000 km/s; known as Moreton waves, they are often observable as broad arcs resembling ripples in a pond.

Corona

Forming the Sun's extended outer atmosphere, the corona is a plasma heated to very high temperatures of 2,000,000°C or more. Scientists remain unsure of the precise cause of coronal heating, but it is thought to be due to a combination of wave heating, magnetic reconnection and microflares. The corona extends out from the Sun in streamers; these merge with the solar wind, an expanding stream of coronal gas blowing through the whole Solar System. Specialist terrestrial observatories atop mountains are capable of securing images of the Sun's corona; it is only visible to the amateur astronomer as a magnificent pearly white structure during a total solar eclipse, when the brilliant photosphere is hidden from view.

Solar projection

Instruments

Pleasing views of the Sun can be obtained with a number of commercially available solar projection devices, which use a small objective lens to collect and focus sunlight onto a small convex mirror inside a shaded box; the reflected light path diverges from the mirror and is projected upon the opposite inner wall of the box. Although they lack the sophistication and versatility of standard telescopic projection, these devices are educational, fun, convenient and safe to use, requiring virtually seconds to set up.

By far the safest way to observe the Sun is to use a small refractor to project its image onto a smooth matt white card which is shaded from direct sunlight. If projecting along the telescope's optical axis, a collar of stiff cardboard can be secured around the telescope tube to provide the shade, allowing more low-contrast detail to be discernable. The white card onto which the Sun's image is to be projected can be held in place with a makeshift frame or a lightweight commercially available

Observing the Sun in white light

Safety

That extreme care must be taken when observing the Sun cannot be overstressed. While light-gathering is at a premium with most types of night-time observation, the Sun is a million times too bright to view directly with the unprotected eye. Unfiltered solar light and heat focused into the eye by any optical instrument – however small and however briefly – will cause eye damage, probable permanent injury to the retina and, at worst, irreversible blindness. One need only think of the burning power of a magnifying glass to realize what concentrated, focused sunlight might do to an unprotected eye.

Never use any so-called "solar filters" that fit over eyepieces. These small disks of dark glass are sometimes supplied with budget telescopes; they offer totally inadequate filtering and are liable to shatter under the intensely hot focused sunlight, allowing direct sunlight to enter the eye. Never use sunglasses, photographic film, smoked glass, reflective or colored confectionery wrappers or a CD as a solar filter to view the Sun, with or without a telescope.

solar projection frame that fastens to the telescope's eyepiece fitting. The distance from the eyepiece to the white card requires adjusting to achieve the right image scale. A solar image projected into a shielded box (with an opening through which the image can be viewed) will enable most detail to be seen.

Refractors of 40–80 mm aperture are commonly used for solar projection. To avoid overheating the eyepiece, larger refractors require a reduction in aperture by employing a mask with a suitably-sized (minimum diameter 40 mm) hole at its center; these are sometimes built into the objective's lens cap. Refractors with relatively long focal lengths of *f*/10 or greater perform best when projecting the Sun's image. Rich-field refractors with very short focal lengths can be used in conjunction with a mask that reduces the amount of light entering the telescope and effectively increases the instrument's focal length. Binoculars attached to a tripod can also be used, but both objective and eye lenses of one half of the binoculars must be securely covered.

It is possible to project the Sun using a Newtonian reflector, but its aperture must be securely fitted with an opaque mask that has a circular hole of no more than 80 mm diameter, positioned off-axis so that direct sunlight is unobstructed by any part of the secondary mirror or its vanes. Large reflectors must never be used at full aperture since the Sun's heat, reflected by the primary mirror onto the much

◀ *A safe means of solar viewing can be achieved by projecting the Sun's image using a small refractor into a box shaded from direct sunlight.*

smaller secondary mirror, can break the secondary mirror or damage its aluminum coating.

Before attempting to use any telescope in this manner, consult the manual to discover if solar observation is contra-indicated. Telescopes of the Schmidt–Cassegrain, Schmidt–Newtonian, Maksutov–Cassegrain or Maksutov–Newtonian design must never be used to project the Sun. Having sealed optics, these types of telescope allow unfiltered sunlight entering them to produce an internal accumulation of heat which could damage the optics and compromise safety. Solar observing with these types of instrument is best performed using proper full-aperture or off-axis solar filters.

Suitable eyepieces

Some budget eyepieces have plastic housings that may melt or fracture with undue heat, or they may have unsuitable cemented lenses that could be permanently damaged if used to observe the Sun. Eyepieces recommended for solar projection are of the inexpensive air-spaced Huygenian or Ramsden type – these eyepieces are often included with inexpensive telescopes, and due to their small fields of view they are frequently discarded without any thought to their usefulness for solar projection. Orthoscopic and Plössl eyepieces have cemented lens elements, and care must be taken when using them for projection.

Locate and focus

Both objective and eye lenses of any finderscopes attached to the main instrument should have their lens caps firmly in place, and no attempt should be made to locate the Sun by sighting it through the finder and centring it in the field. One way of roughly aligning the instrument is to alter the shadow cast by the telescope tube onto the ground so that it appears as small as possible, with the shadow of the tube foreshortened into a circle. Once this is achieved, fit a low-power eyepiece and slew the telescope slightly, using its slow-motion controls, until a bright image of the Sun comes into view on the projection screen. The image will resemble a blurred circle of light, which can be focused using the sharpness of the Sun's limb as a guide. Adjust the distance between the screen and the telescope eyepiece until the Sun's disk is between 100 and 150 mm in diameter.

If the instrument is equatorially mounted, roughly polar-aligned and driven, the Sun will remain in the field of view for a long period of time. An undriven equatorially mounted telescope will require occasional adjustments of the RA control to keep it centered in the field, while continual adjustments in altitude and azimuth will be needed if the instrument is on a simple undriven altazimuth mount.

Herschel wedge

Advanced amateur solar observers can enjoy excellent, full-aperture views of the Sun in unfiltered white light using a Herschel wedge – a specially shaped prism that provides safe observation by reflecting only 5% of the Sun's light into the eyepiece, while diverting 95% of the Sun's energy out of its rear. These devices attach to a telescope in the same manner as a star diagonal. This diverted energy presents a problem in many versions of the Herschel wedge, since the heat is sufficient to damage objects (and people) in its way. Advanced designs, however, use an energy dissipation screen which makes them far safer to use. Viewed through a Herschel wedge, the solar disk appears set against a black sky, and solar features appear sharp and contrasty (providing a good telescope is used in the first place). Herschel wedges require responsible handling and should never be used by anyone who is not completely familiar with their construction and use. They should never be disassembled, nor should any internal filters be removed; if damaged internally, no attempt should be made to use or repair them.

Viewing the Sun through filters

Solar shades

Special solar shades made of high-grade aluminized Mylar filters can be used for viewing the partial stages of solar eclipses or annular eclipse totalities. It is also possible to use them for observing very large sunspots; those with good eyesight are able to discern features as small as one arcminute in diameter (1/30th of the Sun's apparent diameter), such as the disk of Venus during a transit. Although these glasses are usually made out of cardboard with their filters glued in place, they are safe to be used for observing the Sun in a responsible manner. They must never be used if they display signs of wear and tear, and they should be thrown away if there are any large holes or degradation in the filters themselves. Under no circumstances should they be worn while attempting to view the Sun directly through a telescope – the heat of the unfiltered Sun will quickly burn straight through the filters and cause permanent damage to the observer's eyes. Additionally, the shades should never be taken apart and their filters used for makeshift whole-aperture filters for very small telescopes such as finderscopes.

Whole-aperture filters

Solar filters that cover the entire aperture of the telescope (or filters built into an off-axis aperture) allow the observer to view the Sun through the telescope eyepiece, and permit solar imaging using the prime focus, eyepiece projection or afocal methods. There are two main types of solar filter – aluminized Mylar filters and glass Inconel filters. Both types prevent the bulk of the Sun's light and heat from entering the telescope by reflecting it away, permitting only a safe level of solar energy to reach the eye.

Mylar filters

Mylar, a special plastic sheet coated with a film of aluminum, comes in a variety of grades, some of which are suitable for use in solar observing. Numerous everyday commercial products come wrapped in shiny silver Mylar material, and it is even sold as wrapping paper, but this should never be used to construct a makeshift solar filter. Reputable optical suppliers stock the best optical-grade Mylar solar filters, which can be bought premounted in a cell suitable to attach over the end of a variety of instruments, or in individual sheets which can be cut to size. Single-sided Mylar ought to be doubled up to make a suitable filter, while Mylar aluminized on both sides can be used as a single sheet. Photographic density Mylar is unsuitable for use in visual work. Any home-made

▲ *A full-aperture Inconel filter allows safe solar observation through many types of instrument – this shows the author's 200 mm SCT with filter attached.*

Mylar solar filter cell should hold the material securely (but it doesn't need to be stretched taut) and should fit onto the end of the telescope without any risk of its coming loose during the observing session. Direct unfiltered sunlight must not be allowed to enter the telescope tube.

Viewed through a Mylar filter, the Sun appears a shade of blue, since the filter lets in a small amount of light in the blue part of the spectrum. The Sun's color can be changed by inserting a number of different Wratten filters. Since the Sun's heat is not an issue, as it is with the projection method, any good eyepiece may be used to study the filtered Sun; very high magnifications tend to be disappointing because they reveal the effects of atmospheric turbulence. Magnifications ranging from whole-disk views of, say, ×50 to close-up views of interesting features at ×100–150 are possible with any instrument.

Inconel filters

These filters are made of a special metallic coating deposited onto a sheet of optically flat glass. They are more expensive than Mylar filters and their construction makes them rather fragile; handled with care, they last far longer than Mylar filters, which can easily degrade and puncture. Inconel filters are mounted in sturdy metal cells which are custom-sized for a variety of different telescope apertures. They come in two varieties – high density for visual use (such as Thousand Oaks Type 2) and low density for photographic work (Thousand Oaks Type 3). Photographic-grade Inconel filters do not offer adequate filtration to be used safely for visual work. Viewed through an Inconel filter, the Sun appears a pleasant shade of orange – certainly more natural looking than the blue solar disk seen through a Mylar filter. Contrast is good, but the amount of fine solar detail visible tends to be less than that seen through a Mylar filter used on the same instrument under identical conditions.

Conditions for solar observing

Atmospheric turbulence during the daytime tends to rule out viewing detail smaller than 1 arcsecond in diameter, so under average conditions a filtered 100 mm telescope will usually perform as well as a 200 mm telescope. The best observing conditions are found in the early morning when the Sun has risen to an altitude of at least 10° above the horizon. Beyond midday, the ground has usually warmed up so much that turbulence plays havoc with the ability to resolve fine detail. It is best to observe from a grassy area rather than a solid surface, say tarmac or concrete, since grassy surfaces radiate considerably less heat, leading to a more stable image.

Orientation of the solar disk

As the Earth rotates, the Sun's axis appears to rotate from east to west; at sunrise the Sun's axis might be inclined to, or even parallel with the horizon, whereas around midday, when the Sun is due south, the actual orientation of the Sun's axis can be gauged more readily, providing the observer knows the tilt of the Sun's axis with respect to the meridian. Different methods used to view the Sun alter its orientation. Assuming a midday view of the Sun from the northern hemisphere, the Sun's approximate orientation is shown in Table 40:

TABLE 40 ORIENTATION OF THE SUN

Equipment	North	East
Solar shades	Top	Left
Binocular projection	Top	Left
Telescopic projection	Bottom	Left
Filtered telescopic view	Bottom	Right

Top and bottom, left and right should be switched for solar observers in the southern hemisphere.

The Sun's approximate orientation can easily be found by centring the Sun in the field of view and allowing its image to drift; the point of the solar disk that touches the edge of the field will be nearest to its western limb. Centring the Sun's image again and moving the telescope slightly upward, the southern part of the solar limb will first touch the edge of the field (or its northern limb if this is performed in the southern hemisphere).

Twice every year – on January 5 and July 7 – the Sun's axis is aligned with celestial north–south and in line with the meridian when the Sun is due south. The Sun's axis is tilted at its most extreme angle to celestial north–south, by 26.3°, on April 6 and October 10 (see Table 41 showing position angle of the solar north pole during the course of the year with respect to celestial north).

Twice a year – on March 5 and September 8 – the Sun's axis is at its maximum tilt of some 7.2° toward the Earth, and around these times of the year solar features appear to move in curved lines across the solar disk as the Sun rotates. On June 6 and December 7, the Earth lies in the plane of the Sun's equator and neither of its poles are pointed toward us; around these times of the year, solar features appear to move in straight lines across the disk. All of these facts make accurate observations of the Sun something of a challenge to accomplish.

Counting sunspots

A counting sheet of active areas on the Sun's disk is an excellent way to start marking scientific notes of solar activity. A single sheet of ruled paper will be sufficient for a month of observations. At the top of the sheet, list information on the observer, instrument used and the magnification employed, method of observation and total number of observing days in the month. The Carrington rotation numbers of the Sun (which commenced on November 9, 1853) can also be given – these can be found in a variety of almanacs and computer programs. At the side write the days of the month, from 1 to 31, and rule the sheet into several columns, which are to be used for inputting observational data on the following:

1. Time of observation (UT).
2. N – Sun's Northern hemisphere subdivided into two columns headed AA (active areas) and S (spots). If an AA is 10° or more in solar longitude and latitude from another AA it is classed as a separate AA; AAs closer than 10° are classed as a single AA.
3. S – Sun's southern hemisphere subdivided into two columns headed AA (active areas) and S (spots).
4. Total – Subdivided into two columns, AA and S, for the total number of active areas and spots recorded on the Sun.
5. SS Number – The Sunspot Number (R), calculated by using the formula $R = 10g + S$, where g is the number of groups and S is the number of spots. For example, if the total for a session's observation is 5 active areas and 10 sunspots, SS Number (R) = 50 + 10 = 60.

At the bottom of the observing sheet is an additional row listing the totals for each column. Below this, but for the active area columns only, are three boxes in which is written a figure for the mean daily

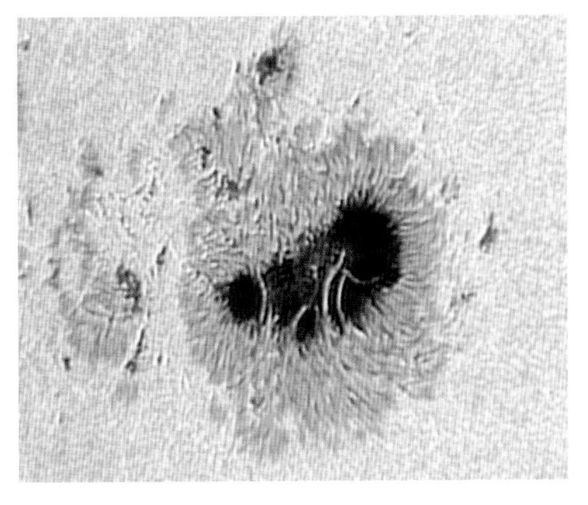

▶ *A detailed image of a sunspot, showing the dark inner umbra, the striated nature of the penumbra and photospheric granulation. The image was taken by Paolo Lazzarotti using a 130 mm refractor, a Herschel prism and an astronomical CCD camera. July 19, 2004, 08:45 UT.*

frequency (MDF), a gauge of the month's solar activity. The MDF is calculated by adding together the number of observing days and dividing the total by this figure. The mean sunspot number (RM) is calculated by adding up the SS Numbers column and dividing it by the number of observing days.

Recording the projected Sun

The diameter of the projected solar disk depends on the telescope's focal length, the focal length of the eyepiece and the distance of the screen from the eyepiece. Whole-disk studies are best made by projecting the Sun's image into a circle of 100, 125 or 150 mm diameter. A special projection blank of 100, 125 or 150 mm diameter, faintly ruled into equal-sized squares identifiable with A–L down one side and 1–12 across the bottom, can be fitted to the screen; this enables the observer to make a drawing of features on a sheet of tracing paper securely held over a similarly ruled blank fastened to a clipboard. The projection screen should be orientated so that the drift of solar features runs parallel to the horizontal lines A–L. This automatically aligns the Sun's projected image with north at the top; with the telescope drive turned off, features appear to drift toward the west.

The drawing should include all of the main features visible on the projection, including detail in and around sunspots and their umbras, plus indications of any faculae visible around the limb regions. A record should be made of the time of the observation and approximate orientation of the Sun's disk – an arrow indicating the direction of drift of the Sun's image across the field of view (with the drive switched off) will prove immensely valuable in calculating the actual orientation of the Sun's axis.

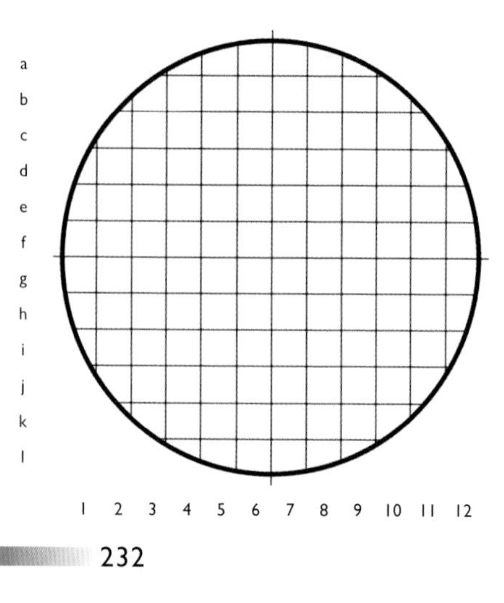

Determining coordinates of solar features

If an accurate observational drawing of active areas and sunspots is secured, and the orientation of the drawing is known, a determination of the actual coordinates of these features on the Sun's surface can be made. A figure known as P gives the position of the Sun's north

◀ *A standard solar projection template allows accurate plots of the Sun's features.*

TABLE 41 POSITION ANGLE (P) OF SOLAR NORTH POLE DURING THE YEAR			
Jan 5, Jul 7	0°	Jul 7, Jan 5	0°
Jan 16, Jun 26	−5°	Jul 18, Dec 26	+5°
Jan 27, Jun 15	−10°	Jul 30, Dec 16	+10°
Feb 8, Jun 2	−15°	Aug 12, Dec 4	+15°
Feb 23, May 18	−20°	Aug 28, Nov 20	+20°
Mar 7, May 7	−23°	Sep 9, Nov 9	+23°
Mar 18, Apr 25	−25°	Sep 21, Oct 29	+25°
Apr 6 (maximum W)	−26.3°	Oct 10 (maximum E)	+26.3°

pole in degrees measured east (+) or west (−) of celestial north throughout the year.

The angle of the Sun's north pole can be marked on an observational drawing by measuring the appropriate number of degrees from north, to the left or right of the north point of the drawing; a line passing through this point, through the center of the disk and on to the other side shows the Sun's meridian.

Stonyhurst disk

As described above (see page 230), the tilt of the Sun's poles with respect to the Earth varies through the year. On June 6 and December 7 the Sun's polar axis is square on to the Earth, and the Earth lies in the Sun's equatorial plane; the Sun's equator makes a straight line across the center of the disk, perpendicular to the polar axis. On March 5 and September 8 the Sun's tilt is at its maximum, and the line of the equator traces its most deeply curved path over the solar disk, the heliographic latitude being displaced by 7.2° S and N, respectively, from the center of the Sun's disk. The value of the Sun's heliographic latitude is known as B_0 and its exact value is given in various ephemerides.

A graphic representation of the changing tilt of the Sun comes in the form of the Stonyhurst disk, which displays lines of latitude and longitude for various values of B_0; a selection of these blanks, with values

TABLE 42 VARIATION OF HELIOGRAPHIC LATITUDE (B_0) DURING THE YEAR			
Dec 7, Jun 6	0°	Jun 6, Dec 7	0°
Dec 16, May 29	1°S	Jun 14, Nov 30	1°N
Dec 23, May 20	2°S	Jun 23, Nov 21	2°N
Jan 1, May 11	3°S	Jul 2, Nov 13	3°N
Jan 10, May 2	4°S	Jul 11, Nov 4	4°N
Jan 20, Apr 21	5°S	Jul 21, Oct 25	5°N
Feb 1, Apr 10	6°S	Aug 3, Oct 12	6°N
Feb 19, Mar 21	7°S	Aug 23, Sep 22	7°N
Mar 5 (maximum S)	7.2°S	Sep 8 (maximum N)	7.2°N

from 0° to 7°, will be sufficient for most solar observers. An observational drawing made on tracing paper, upon which the position of the solar north pole has been marked, can be placed over an appropriate Stonyhurst disk of the same diameter and oriented so that the poles align. The latitude of any of the Sun's features can then be determined by reading its position on the Stonyhurst disk. A number of Internet resources provide downloadable Stonyhurst disk graphics, and copies are also available from the solar observing sections of major astronomical societies.

Determining solar longitude requires an ephemeris or suitable astronomical program; a figure known as L_0 gives the position of the central meridian for the day of the observation. The value of L_0 decreases during the day, and this needs to be allowed for when calculating L_0 for the time of observation. Vertical lines on the Stonyhurst disks represent 10° of solar longitude; perspective means that they get more curved and closer together nearer the Sun's limb. If a sunspot lies 30° to the east (three vertical lines to the right of the central meridian) subtract 30° from the value of L_0 for the day and time of the observation. Calculations of the longitude of features very near the Sun's edge are difficult and less reliable because of foreshortening effects

Solar eclipses

Although the Sun lies 400 times further away than the Moon, it assumes approximately the same apparent angular diameter, as seen from the Earth – around half a degree across – because its diameter is 400 times that of the Moon. Solar eclipses are possible when the center of the Sun is located less than 18.5° from one of the lunar nodes (the point where the Moon's orbital plane intersects the plane of the ecliptic) at the time of new Moon.

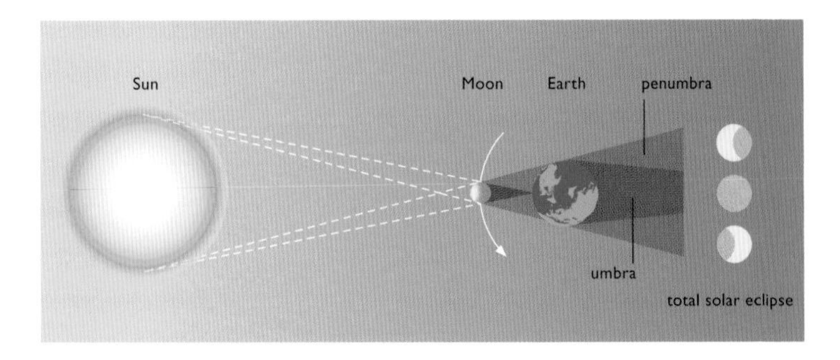

▲ *Sometimes the Moon moves directly between the Sun and the Earth. A solar eclipse will be observed in those areas crossed by the Moon's shadow.*

▲ *Totality, imaged by Robin Scagell from Lizard, Cornwall, on August 11, 1999 – one of the few successful totality images taken from a cloudy southwest peninsula.*

Partial solar eclipses can usually be observed from a wide area of the Earth's surface, and they are not particularly rare from any one location. Seen from near the edge of the Moon's penumbral shadow, only a tiny part of the Sun's edge is hidden by the Moon. Nearer the central line of the eclipse, a greater area of the Sun is obscured by the Moon.

While the Moon's apparent angular diameter varies between 33′ 29″ at perigee and 29′ 23″ at apogee, that of the Sun varies between 32′ 36″ at perihelion and 31′ 32″ at aphelion. Even when the center of the Moon is positioned directly over the center of the Sun, there's no guarantee that the eclipse will be total. On those occasions when the Moon is too small and/or the Sun is too large to produce a total eclipse, an annular eclipse takes place. At mid-eclipse the Moon is surrounded by a spectacular brilliant ring of sunlight.

Where the Moon's dark umbral shadow touches the Earth, a darkened oval forms; viewers within the oval experience a total eclipse. As the Earth rotates beneath the Moon's shadow, a track of totality travels across the Earth's surface. The track's width varies with each eclipse –

it can never be more than 260 km wide, but is often just a few tens of kilometers across. Eclipse chasers travel across the world to be as near to the center of these tracks as possible, in order to experience the maximum duration of totality.

A total solar eclipse is one of nature's most awesome sights. For a brief moment, sadly never more than 7 minutes 40 seconds, and often considerably less, the Sun's brilliant disk is completely hidden by the Moon. Darkness brings with it a distinct chill, and an eerie silence pervades the locality. The brighter stars and planets become visible to the unaided eye. At totality, it is possible to view the eclipsed Sun with the unaided eye – here and there, the edge of the Moon is punctuated by the red prominences of the Sun's chromosphere, and spreading away from the Sun can clearly be seen the pearly streamers of the corona, the Sun's outer atmosphere.

Solar eclipse observation requires extreme care. If you are new to solar observing, please read the safety notice on page 225 before attempting to view the Sun and solar eclipses. By far the best and safest way of viewing a solar eclipse (and, indeed, for solar observing in general) is to project the Sun's image (the technique is described above).

TABLE 43 ANNULAR AND TOTAL SOLAR ECLIPSES 2006–2016

Date	Max eclipse (UT)	Duration	Mag	Type
2006 Mar 29	10:11	4:07	1.052	Total
2006 Sep 22	11:40		0.935	Annular
2008 Feb 7	03:55		0.965	Annular
2008 Aug 1	10:21	2:27	1.039	Total
2009 Jan 26	07:59	6:09	0.928	Annular
2009 Jul 22	02:35	6:39	1.080	Total
2010 Jan 15	07:06	11:08	0.919	Annular
2010 Jul 11	19:33	4:44	1.058	Total
2012 May 20	23:53		0.944	Annular
2012 Nov 13	22:12	2:01	1.050	Total
2013 May 10	00:25		0.954	Annular
2013 Nov 3	12:46	1:40	1.016	Total/annular hybrid
2014 Apr 29	06:03		0.984	Annular
2015 Mar 20	09:46	2:30	1.045	Total
2016 Mar 9	01:57	4:09	1.045	Total
2016 Sep 1	09:07		0.974	Annular

Note: The magnitude represents the maximum fraction of the Sun's disk obscured at mid-eclipse. Values higher than 1.000 mean that the Sun is completely obscured, allowing the chromosphere, prominences and corona to be seen with the unaided eye for a short time.

16 · AURORAE

Among nature's most spectacular phenomena, aurorae take the form of dramatic, dynamic multicolored light displays in the night skies. Occurring between 60 and 400 km high in the Earth's atmosphere, they are caused by the interaction between oxygen and nitrogen molecules and a stream of energetic particles (electrons and protons) from the Sun; these particles are channeled down into the atmosphere by the Earth's strong magnetic field.

Aurorae take on numerous forms, including broad and rayed arcs, flowing curtains and fabulous streaming coronae, all of which may appear to change in color, form and intensity as the activity progresses. Red auroral displays take place high in the atmosphere and result from the impact of solar particles with oxygen. Lower down in the atmosphere, at heights of around 100 km, the impact of solar particles with oxygen is more frequent and a soft green color is produced. Pink colors visible at the base of bright surging auroral arcs are caused by nitrogen, which glows blue at higher altitudes.

Low levels of auroral activity occur continually in the atmosphere surrounding the Earth's geomagnetic poles, at geomagnetic latitudes of 67° and higher. Increased activity on the Sun produces heightened levels of auroral activity, which can be visible at much lower latitudes. Consequently, the frequency of major auroral activity reflects the

▲ *Panoramic image of an aurora, captured by Peter Lawrence from Selsey on the English south coast, in the early hours of October 30, 2003.*

11-year solar cycle, with a peak in frequency between one and two years following solar maximum. Auroral displays can be observed as far south as the Mediterranean or as far north as southern Australia several times during every solar cycle. Great displays are visible in tropical locations only a few times per century, recent major aurorae having taken place on March 13–14, 1989, and during October and November 2003.

In the northern hemisphere the phenomenon is referred to as the aurora borealis; in the south, it is known as the aurora australis. Both are caused by identical mechanisms; they may be considered as two sides of the same coin, since auroral activity in the northern hemisphere is mirrored by activity in the southern hemisphere, although its visibility over either hemisphere depends on the date and time of its occurrence. For example, the southern counterpart of a magnificent auroral display visible in the night skies of large parts of Western Europe during the winter will not be visible from Australia, which is basking in the summer sunshine at that time.

Observing aurorae

Light pollution is detrimental to many forms of astronomical observation, and observing aurorae is dependent on having a reasonably dark, clear sky. Urban skyglow is capable of drowning out the visibility of most aurorae, although the brighter components of major displays can sometimes be glimpsed from the city. A small hand-held spectroscope with its slit open can discriminate between various forms of skyglow and true aurorae, since it will reveal the green auroral emission line of oxygen.

Observing aurorae requires little equipment other than a keen pair of eyes; events can be recorded in a notepad illuminated by a red torch, and timings can be made with an ordinary watch. Photographs of aurorae can be made with a standard 35 mm camera on a tripod, the time exposure depending on the brightness of the event. Digital cameras are capable of imaging aurorae at much shorter exposures than conventional photography, minimizing star trails and capturing the moving forms of the aurora more sharply.

Location

Latitudes further north (or further south, if in the southern hemisphere) will generally be more favorable in terms of increased frequency of auroral visibility. However, the auroral zones of activity are centered around the geomagnetic poles, so they are offset from the geographical poles by a considerable degree. The center of the northern auroral zone currently lies over 79° N, 69° W, near Thule in Greenland,

▶ *The Moon shines beneath the glowing curtain of the aurora borealis, in this image by Jamie Cooper.*

so observers in North America are treated to a greater frequency of aurorae than are observers in Europe and Asia at the same geographical latitudes. Southern hemisphere auroral activity takes place in an oval centered over the southern magnetic pole, which lies near the Antarctic coast around 2500 km from the geographical South Pole. Favored locations for viewing the aurora australis lie across a great swathe of Antarctica, including the South Pole, but auroral displays can also be observed from New Zealand and southern Australia.

Aurorae can be observed from any point on the Earth's night-time surface that lies within the zone of auroral activity. Aurorae will appear at their highest at local magnetic midnight, when the Sun, the magnetic pole and the observer are all in line. From the UK this occurs at around 22h UT; from New York, USA, 05h UT; from Seattle, USA, 10h UT; from Wellington, New Zealand, 14h UT; from Perth, Australia, at around 17h UT.

Because of the greater altitude of red displays they can be observed from lower latitudes. Aurorae viewed in dusk or dawn skies in the late spring and summer may display a vivid violet hue because they selectively reflect the violet wavelengths of sunlight.

Auroral development

Low-level quiescent aurorae, associated with holes in the solar corona, are often visible in the skies on several successive evenings, appearing as a fairly unremarkable homogenous glow near the horizon. There is usually advance warning of impending major auroral activity because the vast quantities of highly energetic particles spewed out by solar flares and coronal mass ejections can take around 2 to 4 days to reach the Earth, giving observers time to gear up for observing and/or recording these events.

A number of commonly observed auroral forms characterize the development of activity. A display often begins with a homogenous glow

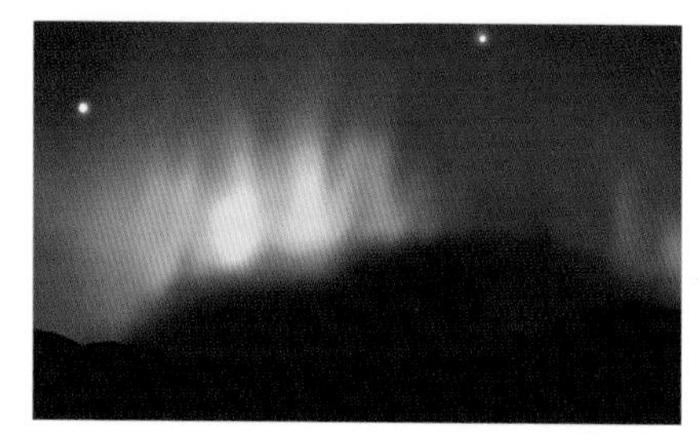

◀ *From top to bottom: a homogenous auroral glow; an auroral arc; a rayed arc; a rayed band. Graphics by the author.*

near the horizon, which develops into clearly defined arcs of light; numerous linear beams may project upward from them. If activity continues to develop, the aurora may extend to latitudes further south of the observer's location, producing a remarkable auroral form known as a corona, with rays that converge to a point overhead. Coronal forms are produced by a perspective effect, as the observer peers upward into enormously high vertical rays. A veil of light may cover large portions of the sky, forming a background for the more defined activity. The phenomenon declines into patches of light as activity subsides. The entire process may be repeated again on a single evening if the storm is severe.

Recording aurorae

In order to record the appearance, development, structure and intensity of auroral forms, a standard observational shorthand is commonly used on report forms. Annotated sketches showing an aurora's development are also useful to include in any report.

Elevation and direction

Observers should note the elevation of the activity above the horizon, from its base to its highest point; estimations will do, but more accurate measurements can be made using a simple home-made sighting device such as an alidade. A compass can be used to determine the azimuth direction of activity.

Condition of activity

q – Quiet. No variations observed in form or intensity.

Active conditions (a)
a1 – Active. Folding or unfolding of bands.
a2 – Active. Rapid changes in the shape of the auroral form's lower edges.
a3 – Active. Rapid horizontal motion (to west or east) of rays.
a4 – Active. Forms rapidly fading, quickly replaced by others.

Pulsating conditions (p)
p1 – Pulsating. Rhythmic, uniform changes in brightness.
p2 – Pulsating. Flaming, surges of light rippling upward through the auroral forms.
p3 – Pulsating. Flickering, with rapid irregular variations in brightness.
p4 – Pulsating. Streaming, irregular changes in brightness observed to ripple through homogenous forms.

Auroral structure

N – Glow. A generally undefined luminosity near the horizon. Glows along the northern horizon (or southern horizon in the southern hemisphere) are the first indications of auroral activity. They may be difficult to distinguish from the light of twilight or urban glows, but an observer thoroughly familiar with his or her observing site will usually be able to tell the difference.

A – Arc. A broad curving arch, often pale green or white. Arcs become visible above the northern horizon (or southern horizon in the southern hemisphere) as the aurora develops into more distinct forms and pushes toward the observer, decreasing in latitude.

R – Ray. A searchlight-like beam of light. They can appear in conjunction with arcs and bands, or in complete isolation, pointing upward into the sky like a searchlight.

RA – Rayed Arc. An arc from which rays extend upward. They often develop when an arc's brightness suddenly increases.

B – Band. An arc that has developed into twisted or folded band forms, like streaming ribbons.

RB – Rayed Band. A band from which rays extend upward.

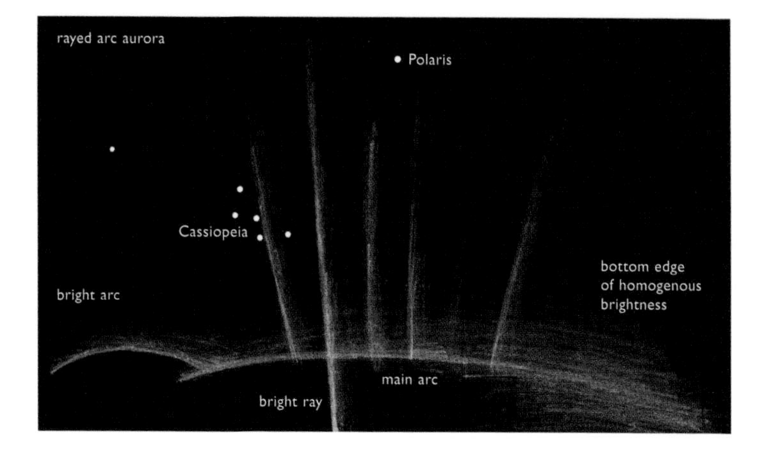

▲ *An observational drawing by the author of the rayed arc stage of a particularly fast-moving auroral display. It was observed on February 8, 1986, between 22:50 and 23:15 UT from the English midlands.*

P – Patch. A diffuse, ill-defined cloud of light, often visible as activity declines.

V – Veil. A diffuse glow covering a large area of sky, usually white.

C – Corona. Rays emanating from a single point near the zenith, visible on rare occasions when an aurora pushes into the southern half of the observer's sky. Coronal formations often appear shortly before activity begins to decline.

Qualifying symbol

m – Multiple, with several groups of the same type of form (for example, m4R describes four rays).
f – Fragmentary. Part of an arc or band.
c – Coronal, with patches or rays converging overhead.

Structure of forms

H – Homogenous, lacking in structure and usually of uniform brightness.
S – Striated, with mixed bright and dark horizontal lines.
R1 – Rayed, with short rays up to 20° long.
R2 – Rayed, with medium length rays between 20° and 60° long.
R3 – Rayed, with rays longer than 60°.

Brightness

1 – Faint, of similar brightness to the Milky Way, too dim to perceive its color.
2 – As bright as moonlit cirrus cloud.
3 – As bright as moonlit cumulus cloud.
4 – Bright enough to easily read by and to cast a shadow.

Color

a – Red in upper part of display, green in lower part.
b – Red lower border.
c – White, pale green or yellow.
d – Red in main part of auroral form.
e – Mixed red and green.
f – Blue and purple.

The layout of aurora observing forms differs between astronomical societies, but all use the above system of shorthand, which is input in the appropriate places on the observing form.

GLOSSARY

Albedo A measure of an object's reflectivity. A pure white reflecting surface has an albedo of 1.0 (100%). A pitch black, non-reflecting surface has an albedo of 0.0. For example, the Moon is a dark object with an overall albedo of 0.07 (reflecting 7% of the sunlight that falls upon it). Venus is highly reflective, its clouds having an albedo of 0.65.

Altitude The angle of an object above the observer's horizon. An object on the horizon has an altitude of 0°, while at the zenith its altitude is 90°.

Aperture The diameter of a telescope's objective lens or primary mirror.

Aphelion The point in an object's orbit furthest from the Sun.

Apogee The point in an object's orbit where it is furthest from the Earth. At apogee, the Moon can reach a maximum distance of 406,700 km from the Earth.

Apparition The period of time during which a planet, asteroid or comet can be observed. For example, apparitions of Mercury last only a few weeks, whereas those of Jupiter last for many months.

Arcminute One minute of arc. 1/60th of a degree. Indicated with the symbol ′.

Arcsecond One second of arc. 1/60th of an arcminute. Indicated with the symbol ″.

Asteroid A minor planet. A large solid body of rock in orbit around the Sun.

Astronomical Unit A convenient measure of distances within the Solar System. It is based on the average distance of the Earth from the Sun. 1 AU is equal to 149,597,870 km.

Atmosphere The mixture of gases surrounding a planet, satellite or star.

Aurora The glow caused by excitation of gases in a planet's atmosphere; it is produced by energetic solar particles channeled by the planet's magnetic field.

Axis The imaginary line around which a body rotates.

Basin A very large circular structure, usually formed by impact and comprising multiple concentric rings.

Caldera A sizable depression in the summit of a volcano, caused by subsidence or explosion.

Catena (plural catenae) A chain of craters.

Central peak An elevation found at the center of an impact crater, usually formed by crustal rebound after impact.

Cloud belt/zone A dark/bright region, lying parallel to the planet's equator, in the atmosphere of any of the four gas giants.

Comet A relatively small, solid nucleus of rock and ice in orbit around the Sun. When a cometary nucleus approaches the Sun it heats up and its ices sublimate, giving rise to a large coma around the nucleus and longer tails of gas and/or dust.

Conjunction The apparent close approach of a planet to the Sun or another planet, seen from the Earth. A planet is in conjunction with the Sun when the Sun lies between that planet and the Earth. Venus and Mercury also undergo inferior conjunction when they lie directly between the Earth and the Sun, but they cannot be seen at these times unless they are in transit across the Sun's face.

Crater A circular feature, often depressed beneath its surroundings, bounded by a circular (or near-circular) wall. Almost all of the large craters visible in the Solar System have been formed by asteroidal impact, but a few smaller craters are endogenic, of volcanic origin.

Crescent The phase of a planet or other spherical body seen when it is observed to be between 0 and 50% illuminated.

Culmination The passage of a celestial object across the observer's meridian, when it is at its highest above the horizon.

Dark side The hemisphere of a solid body not experiencing direct sunlight.

Degree As a measurement of an angle, one degree is 1/360th of a circle. Indicated by the symbol °. In terms of heat, degrees are increments of a temperature scale. Scales most commonly used in astronomy are Celsius (C) and Kelvin (K). 0°C, or 273.16 K, is the freezing point of water. 0 K, or −273.16°C, is known as absolute zero, the temperature at which all molecular movement ceases.

Dichotomy Half-phase of a planet or satellite, when it is observed to be exactly 50% illuminated.

Dome A low, rounded elevation with shallow-angled sides. It can be formed volcanically or through subcrustal pressure.

Earthshine The faint blue-tinted glow of the Moon's unilluminated hemisphere, visible with the naked eye when the Moon is a narrow crescent. It is caused by sunlight reflected onto the Moon by the Earth.

Eccentricity A measure of how an object's orbit deviates from circular. A circular orbit has zero eccentricity. Eccentricity between 0 and 1 represents an elliptical orbit.

Ecliptic The apparent path of the Sun on the celestial sphere during the year. The ecliptic is inclined by 23.5° to the celestial equator. The major planets follow paths close to the ecliptic, and the Moon's path is inclined by some 5° to it.

Ejecta A sheet of material thrown out from the site of a meteoroidal or asteroidal impact that lands on the surrounding terrain. Large impacts produce ejecta sheets composed of melted rock and larger solid fragments, in some cases producing bright ray systems.

Elongation The apparent angular distance of an object from the Sun, measured between 0 to 180° east or west of the Sun. For example, the first quarter Moon has an eastern elongation of 90°; Venus has a maximum possible elongation of 47°.

Ephemeris A table of numerical data or graphs that gives information about a celestial body in a date-ordered sequence, such as the rising and setting times of the Moon, the changing illumination of Venus, the longitude of Jupiter's central meridian, and so on.

Equator The great circle of a celestial body whose plane passes through the body's center and lies perpendicular to its axis of rotation.

Fault A crack in the crust of a solid object caused by tension, compression or sideways movement.

First quarter Half-phase between new Moon and full, occurring one-quarter the way through the lunation.

Full Moon When the lunar disk is observed to be completely illuminated by the Sun. Viewed from above, the Sun, Earth and Moon are in line, with the Sun and Moon on opposite sides of the Earth.

Galilean moons/satellites The four largest satellites of Jupiter, namely Io, Europa, Ganymede and Callisto.

Gas giants The four largest planets in the Solar System – Jupiter, Saturn, Uranus and Neptune – each of which has a thick gaseous atmosphere and no visible solid surface.

Gibbous The phase of a spherical body between dichotomy (50% illuminated) and full (100% illuminated).

Graben A valley bounded by two parallel faults, caused by crustal tension.

Highlands Heavily cratered or mountainous regions of the Moon or planets.

Impact crater An explosive excavation in the crust of a solid body formed by a large projectile striking at high speed.

Jovian Relating to Jupiter.

Lava Molten rock extruded onto the surface of a body by a volcano.

Limb The edge of the Sun, Moon or planet.

Lithosphere The solid crust of the Moon or terrestrial planet.

Lunar Pertaining to the Moon (from Luna, Roman goddess of the Moon).

Lunar eclipse A period during which the Moon moves through the shadow of the Earth. Lunar eclipses can be penumbral, partial or total. They happen at full Moon, when the Sun, Earth and Moon are almost exactly in line.

Lunar geology The study of the lunar rocks and the processes that sculpted the Moon's surface; often referred to as selenology.

Lunation The period taken for the Moon to complete one cycle of phases, from new Moon to new Moon, averaging 29 days, 12 hours and 44 minutes. This is the Moon's synodic month. Lunations are numbered in sequence from Lunation 1, which commenced on January 16, 1923. Lunation 1000 commenced on October 25, 2003.

Mare (Latin sea; plural maria) A large, dark lunar plain. Maria fill many of the Moon's large multiringed basins and comprise a total of 17% of the Moon's entire surface area.

Massif A large mountainous elevation, usually a group of mountains.

Meteorite A meteoroid that has survived a passage through the Earth's atmosphere.

Meteoroid A small solid body composed of rock or metal in orbit around the Sun. Meteor streams encountered by the Earth were deposited by comets, and produce the annual meteor showers.

Mons (Latin mountain; plural montes) The generic term for a mountain on the Moon or terrestrial planet.

New Moon The lunar phase during which all of the near-side is unilluminated. Seen from above, the Moon lies directly between the Earth and Sun.

Occultation The disappearance or reappearance of a star or planet behind the limb of the Moon or another planet.

Opposition The position of a planet when its celestial longitude, measured from the Earth, is 180° to that of the Sun; at opposition, a planet appears to be opposite to the Sun and transits the meridian at midnight. Opposition also coincides with the point in a planet's apparition when it is nearest to the Earth.

Perigee The point in an object's orbit where it is closest to Earth. At perigee, the Moon can be as close as 356,400 km to the Earth.

Perihelion The point in an object's orbit when it is nearest the Sun.

Phase The degree to which a planet or satellite is observed to be illuminated by the Sun. Phases can be crescent (less than 50% illuminated) or gibbous (more than 50% illuminated).

Planet One of nine large objects in orbit around the Sun, ranging from small, solid Pluto (2274 km across) to the large gas giant Jupiter (142,984 km across).

Quadrature The position of a planet when it has an elongation of 90° from the Sun.

Ray A bright (though sometimes dark) streamer of material radiating from an impact crater.

Rift valley A graben-type feature caused by crustal tension, faulting and horizontal slippage of the middle crustal block.

Rille A narrow valley. Some rilles are linear, caused by crustal tension and faulting; others are sinuous, believed to have been produced by fast-moving lava flows.

Rings Material distributed in a narrow plane above the equators of all four gas giants, Jupiter, Saturn, Neptune and Uranus.

Satellite A small body revolving around a larger body.

Secondary cratering Craters produced by the impact of pieces of solid debris thrown out by a large impact. Secondary craters often occur in distinct chains, where piles of material impacted simultaneously.

Seeing A measure of the quality and steadiness of the atmosphere as it affects an image seen through the telescope eyepiece. The quality of seeing is affected by atmospheric turbulence, caused largely by thermal effects.

Solar Relating to the Sun.

Solar System The Sun and everything within its gravitational domain.

Sun The central star of the Solar System.

Terminator The line separating the illuminated and unilluminated hemispheres of a planet or satellite.

TLP Transient Lunar Phenomena. A rarely observed, short-lived anomalous colored glow, flash or obscuration of local surface detail; the causes are poorly understood.

Transit The passage of a celestial object across the face of another, larger celestial object. For example, Venus occasionally transits the Sun, and Io frequently transits Jupiter.

Universal Time (UT) The standard measurement of time used by astronomers over the world. UT is the same as Greenwich Mean Time, and it differs from local time according to the observer's position on the Earth and the time conventions adopted in that country.

Volcano An elevated feature built up over time by the eruption of molten lava and ash.

Zenith The point in the sky directly above the observer.

RESOURCES

Societies

American Association of Amateur Astronomers
P.O. Box 7981
Dallas, TX
75209-0981 U.S.
Website: www.astromax.com

The AAAA website includes an online newsletter, astronomical guidebook, planetary data, images and links.

Association of Lunar and Planetary Observers (ALPO)
Website: www.lpl.arizona.edu/alpo

Based in the United States, ALPO is organized into observing sections, each dedicated to its own planet or other object/phenomenon and headed by an observing coordinator. The ALPO journal publishes data throughout the year. ALPO holds an annual conference as a means for members to gather and formally present their observations, research and findings in person.

Astronomical League
9201 Ward Parkway, Suite 100
Kansas City, MO
64114 U.S.
Website: www.astroleague.org

The Astronomical League is composed of more than 240 local organizations from across the United States. Contact them to find a comprehensive list of amateur astronomy groups.

Lunar and Planetary Institute (LPI)
Website: www.lpi.usra.edu

LPI, based in Houston, Texas, is a focus for academic participation in studies of the current state, evolution and formation of the solar system. It has extensive collections of lunar and planetary data, an image-processing facility, an extensive library, education and public outreach programs, resources and products. The LPI also offers publishing services and facilities for workshops and conferences.

Royal Astronomical Society of Canada
136 Dupont Street
Toronto, ON
M5R 1V2 Canada
Website: www.rasc.ca

The principal astronomical society in Canada, its membership includes amateur and professional astronomers from around the world.

Notable websites – general astronomy and space interest

Astronomy Picture of the Day
Website: antwrp.gsfc.nasa.gov/apod/astropix.html

Each day a different image or photograph of our fascinating universe is featured, along with a brief explanation written by a professional astronomer.

Eric Weisstein's World of Astronomy
Website: scienceworld.wolfram.com/astronomy

One of the Internet's best fully searchable resources for astronomical information, data and definitions.

Let's Talk Stars
Website: www.letstalkstars.com/listennow.html

Starizona's archive of astronomy radio broadcasts by David and Wendee Levy.

The Nine Planets
Website: www.nineplanets.org

A fascinating and well-researched multimedia tour of the solar system.

The Sky at Night
Website: www.bbc.co.uk/science/space/spaceguide/skyatnight/

Features current astronomical events and information, plus a video archive of previously broadcast Sky at Night programs.

Solarviews
Website: www.solarviews.com

A vivid multimedia adventure unfolding the splendor of the sun, planets, moons, comets, asteroids and more.

Notable websites – observing resources

American Meteor Society
Website: www.amsmeteors.org

A great resource for information on observing meteors.

Comet Observation Home Page
Website: encke.jpl.nasa.gov

Excellent resource for cometary observers and researchers, with the latest images and ephemerides.

Eclipses Online
Website: www.eclipse.org.uk

In conjunction with the U.S. Naval Observatory, a canon of eclipses based on software used in the production of The Astronomical Almanac. It provides global and local circumstances, animations and eclipse panoramas for partial, annular, total and hybrid solar eclipses in the period AD 1501 to 2100. Global circumstances of penumbral, partial and total lunar eclipses are available for the same period. Most of the information can be downloaded.

The International Meteor Organization
Website: www.imo.net

With 250 members from around the world the IMO is an excellent resource for the study of meteor showers and their relation to comets and interplanetary dust.

Mercury Chaser's Calculator
Website: www.fourmilab.ch/images/3planets/elongation.html

Displays the date, time and distance of the maximum elongations of Mercury for a given year.

SOHO
Website: sohowww.nascom.nasa.gov

The orbiting Solar and Heliospheric Observatory gives a variety of updated views of the sun in various wavelengths, plus current information on geomagnetic activity and auroras.

Solar System Live
Website: www.fourmilab.ch/cgi-bin/uncgi/Solar/action?sys=-Sf

An interactive orrery showing the planetary positions for any date and time.

U.S. Naval Observatory
Website: aa.usno.navy.mil/data

Ephemerides for a great deal of astronomical information useful in planning or researching observations.

Astronomy magazines

Amateur Astronomy
Website: www.amateurastronomy.com

A quarterly journal with articles written for, by and about amatuer astronomers from around the world

The Astronomer
Website: www.theastronomer.org

Monthly newsletter for observers. TA, as it is known, has been published since 1964 with the aim of publishiing all observations of astronomical interest as soon as possible after they are made.

Astronomy
Website: www.astronomy.com

Monthly general-interest astronomy magazine, with the latest news and events.

Astronomy Now
Website: www.astronomynow.com

Monthly magazine featuring space news, observing articles and information.

Mercury Magazine
Website: www.astrosociety.org/pubs/mercury/mercury.html

Bi-monthly magazine for members of the Astronomical Society of the Pacific.

Night Sky
Website: nightskymag.com

Monthly magazine aimed at astronomy beginners.

Popular Astronomy
Website: www.popastro.com/spapop/home.htm

Quarterly magazine published by the Society for Popular Astronomy, featuring articles, observations and a full sky diary.

Sky at Night
Website: www.skyatnightmagazine.com

Monthly magazine featuring cover CD-ROM. Contains guides to equipment, observing notes.

Sky and Telescope
Website: skyandtelescope.com

Monthly general-interest magazine featuring news items and observing notes.

INDEX

General index

See also object index
Page numbers in *italics* refer to illustrations or their captions.

Object index

See also general index
Page numbers in *italics* refer to illustrations or their captions.

ACKNOWLEDGMENTS

Frances Button of Philip's has provided all the help any author could ever wish for, from the book's beginning through to its completion – thank you!

My thanks to Britain's Society for Popular Astronomy, many of whose helpful officers, directors, advisors and ordinary members have provided help and encouragement with observing and research, not only for this book, but throughout the 23 years of my membership. Grischa Hahn and Hans-Joerg Mettig (jupos.org) and Damian Peach helped with the section on Jupiter observation.

I am indebted to the following people for providing many of the splendid observations and images in this book: Anthony Ayiomamitis, Richard Bailey, Mike Brown, Jamie Cooper, John Fletcher, Brian Jeffrey, Stefan Lammel, Pete Lawrence, Paolo Lazzarotti, Nigel Longshaw, Cliff Meredith, Damian Peach, Richard Prettyman, Robin Scagell/Galaxy Picture Library, Kevin Smith, Paul Sutherland, Nik Szymanek, Dave Tyler, Sean Walker and Grahame Wheatley.

Finally, the successful completion of this book would not have been possible without the limitless patience, incredible help and immense support given my wife Tina and my daughter Jacy.

Picture credits

Anthony Ayiomamitis 198; Richard Bailey 221r, 223; Mike Brown 115, 126, 142, 191, 193, 196t; John Chapman-Smith/Galaxy 226; Jamie Cooper 7, 9, 14, 15t, 15b, 16, 17, 18, 30, 32b, 52, 65, 66, 69, 80, 196b, 199, 200, 239; ESA 163; John Fletcher 102, 138, 155, 160; Peter Grego 27, 41, 44, 45, 46, 56, 62, 68, 75, 81, 82, 83, 84, 85, 86, 87, 88, 89, 93, 100, 108, 109, 113, 117, 118, 121, 122, 135, 137, 146, 149t, 149b, 157, 166, 168, 171, 173, 174, 180, 183, 184, 187, 188, 192, 203, 207, 221l, 228, 240, 242; Brian Jeffrey 197b; Stefan Lammel 31; Pete Lawrence 67, 175l, 186, 190, 211, 237; Paolo Lazzarotti 231; Nigel Longshaw 195; Cliff Meredith 124; NASA/JPL 49, 58, 73, 74, 76, 107b, 154; NASA/JPL/DLR 106b, 107t; NASA/JPL/Lunar and Planetary Laboratory 106t; NASA/JPL/Malin Space Science Systems 78; NASA/STScI 94, 140, 150, 159t, 159b; NASA/JPL/Univ. Arizona 1, 104; Richard Prettyman 175r; Robin Scagell/Galaxy 164, 235; Courtesy SOHO/EIT Consortium 214; Dave Tyler 105, 130, 144, 151, 197t, 220; Sean Walker 222; Grahame Wheatley 194, 206.

Artwork

Philip's 4, 10, 12, 19, 25, 29, 32, 35, 39, 43, 96, 165, 177, 189, 209, 218, 234; Peter Grego/Philip's 5, 11, 50, 51, 53, 60, 63, 64, 70, 79, 125, 132, 161, 215, 232.